T0283314

THE
FORGOTTEN
SENSE

THE
FORGOTTEN
SENSE

The New Science of Smell—
and the Extraordinary Power of the Nose

JONAS OLOFSSON

MARINER BOOKS
New York   Boston

THE FORGOTTEN SENSE. Copyright © 2023 by Jonas Olofsson. Translation copyright © 2025 by Jonas Olofsson. All rights reserved. Printed in the United States of America. No part of this book may be used or reproduced in any manner whatsoever without written permission except in the case of brief quotations embodied in critical articles and reviews. For information, address HarperCollins Publishers, 195 Broadway, New York, NY 10007.

HarperCollins books may be purchased for educational, business, or sales promotional use. For information, please email the Special Markets Department at SPsales@harpercollins.com.

The Mariner flag design is a registered trademark of HarperCollins Publishers LLC.

Originally published as *Det underskattade sinnet* in Sweden in 2023 by Natur & Kultur.

FIRST US EDITION

Designed by Emily Snyder

Library of Congress Cataloging-in-Publication Data

Names: Olofsson, Jonas (Professor of psychology), author.
Title: The forgotten sense : the new science of smell—and the extraordinary power of the nose / Jonas Olofsson.
Other titles: Det underskattade sinnet. English
Description: First US edition. | New York : Mariner Books, [2025] | Includes bibliographical references and index.
Identifiers: LCCN 2024023833 (print) | LCCN 2024023834 (ebook) | ISBN 9780063394674 (hardcover) | ISBN 9780063394704 (ebook)
Subjects: LCSH: Smell. | Olfactory nerve. | Nose.
Classification: LCC QP458 .O456 2025 (print) | LCC QP458 (ebook) | DDC 612.8/6—dc23/eng/20241001
LC record available at https://lccn.loc.gov/2024023833
LC ebook record available at https://lccn.loc.gov/2024023834

ISBN 978-0-06-339467-4

24 25 26 27 28 LBC 5 4 3 2 1

CONTENTS

INTRODUCTION

THE BEST WORST SMELL IN THE WORLD

I N VÄSTERBOTTEN, A northern province of Sweden, there is a special light in the late summer evenings. The sun is setting, but the sky is still bright with this cool light that reflects the deep blue hues of the waters of the Gulf of Bothnia and a rain cloud coming in from the east. In the west, the shimmering sunset glows in pink and orange tones. Added to this are the saturated greens and browns of the birch, pine and spruce trees around us. The light and colors are not only a backdrop, but for me they are also necessary ingredients for a traditional Västerbotten surströmming feast. Tonight is the premiere. The tin can of herring is placed on a platter with juniper rice on the veranda table. The can bulges out in all directions, bursting with perhaps the world's most controversial smell. The brand is Oskars, the preferred brand here in the region, which has been sold out in all stores in recent weeks. Oskars herring is distinguished, its supporters say, by its firm shape; it does not fall apart despite the intense fermentation process to which it has been subjected. This facilitates the gutting and filleting of the fish, which is complicated and inevitably leaves your hands smelling for days, no matter how well you wash them. Today's jar, purchased several weeks ago, has had plenty of time to ferment. This is not a product that decays immediately. On the contrary, it is as if the ravages of time do not affect the herring at all. Like a character in a Torgny Lindgren novel, the herring has reversed the aging process. It does not get worse with time, but better. For the truly inveterate sour-herring lover, only last year's herring can provide total satisfaction.

Sour herring is a fragrant delicacy for many of us from Norrland. But its distinctive smell and taste are not fully appreciated even in my own family. Sour herring divides our table. I, my mother and my sister sit on the herring side of the table. Our respective partners and children, who come from Stockholm and the United States, sit on the other side, where an alder-chip-smoked side of salmon is the main course. Two related yet completely different smells mingle at the table.

What makes the smell of herring so polarizing? As a professor and researcher on the sense of smell, I can't help but conduct little experiments at the dinner table. My son Henry, two months old, doesn't seem to mind the smell of herring at all, even though I hold a piece directly under his nose. We humans are born with a fully developed sense of smell, and scientists have discovered that babies can learn to like smells in the womb, so they are attracted to what their mothers ate during pregnancy. This is Henry's first exposure to surströmming—his American mother would never willingly eat it— but perhaps tonight's strong smells will influence his sense of smell for the rest of his life.

Someone who is not so neutral is my niece Ellen, who declares with five-year-old assertiveness that first, she doesn't want to eat herring, and second, she doesn't want to smell it! Then she storms into the cottage with her blond curls following close behind. She is excused. At the age of five, children have already learned what smells are considered pleasant and unpleasant, and sour herring is not on offer at her preschool in a Stockholm suburb. As all parents of young children know, five-year-olds often have strong preferences. So strong that it takes scientific methods and a lot of patience to get them to like new tastes and smells. I take this defeat in stride, and hope that over the years she may change her mind.

Tonight's third subject is not as articulate as Ellen, but has the best nose in the family. Nelson the miniature poodle is magnetically attracted to strong smells and is known to eat anything that comes his way. His favorite game is sniffing for dog treats hidden around the apartment. I ask him to sit nicely and hold out two small pieces of fish, salmon in my left hand and sour herring in my right. It's a simple test to see which smell he prefers. He sniffs a few times before making his choice: the salmon. But then he also eats the surströmming. With some goodwill, I can call it a draw.

Surströmming's global reputation has probably been shaped by the many YouTube videos of people trying it for the first time. The results are predictable: terrified faces and wrinkled noses when the can is opened, hesitation and fear of the fish itself, and gagging when eating it. The creators of such content have rarely taken the trouble to prepare the fish in a traditional way, with the right accessories. Not to mention the light—it is far from that of Västerbotten. No wonder the herring is a flop in such circumstances.

Surströmming is not one of Sweden's major export products, and is likely to remain obscure. My own attempts to spread its gospel to the US have failed. Fermented herring is strictly forbidden for air travel—the jar is classified as an explosive material. The thought of being labeled a terrorist is a deterrent. Ordering from Europe to the US is a safer option, but the prices are outrageous. So I have not yet had the pleasure of treating my American friends and family to this northern gold. However, when we have invited friends and acquaintances to a surströmming party in Stockholm, the interest has been great. They come from all corners of the world, and have only heard about the mythical fish. But interest usually fades the moment the jar is opened, and only the bravest take a bite. Most of the herring ends up in my own stomach. That is fine by me.

But even in the heartland of surströmming, it's hard to get a full commitment to it. We surströmming enthusiasts do our best to keep the family tradition alive; we comment and discuss exactly how much of each ingredient is needed—the chopped onion, the boiled potatoes, the tomatoes, the sour cream, the butter on the flatbread—and in what order they should be applied. We agree that the herring should be cleaned by hand, and that pre-filleted herring is somehow suspect, something that suits Stockholmers and other amateurs. It is more of a ritual than a dish. The smelly fish always leads to discussions, and they continue well into the northern summer evening. Only once is there silence, when my brother-in-law asks the not entirely innocent question: If the herring is as good as you say, why do you only eat it once a year? I answer with a quote from Ingemar Stenmark, the laconic downhill-skiing legend: "It's not worth explaining to those who don't understand." Then we change the subject.

Smelling is the easiest and most natural thing in the world. We do it all the time. We breathe in every four seconds. That's up to one thousand inhalations per hour and over twenty thousand times per day. The air we breathe in and out carries vital oxygen, which is sucked down our windpipes and into the alveoli, where it is transferred to the bloodstream and can oxygenate all parts of the body. But the air contains more than oxygen, and on its way to and from the lungs it passes through the nose, mouth and throat. There, a number of remarkable biological processes take place that shape how you experience the world. I and my colleagues at Stockholm University explore the vastness of the sense of smell—how smells create experiences, emotions, memories, behaviors. My research group consists of neuroscientists, psychologists, linguists, sensory scientists and computer scientists who share a common interest in the sense of smell. Nowadays, the human sense of smell is a popular research topic, but for many years it was almost completely unexplored. It was neglected and devalued by many philosophers and scientists. Moreover, until recently, odors were difficult to assess and describe with concepts and objective measuring instruments. A myth emerged that the human sense of smell is weak and insignificant. Therefore, the sense of smell still holds many mysteries waiting to be explored. This creates a pioneering spirit among us olfactory researchers.

The sense of smell is fascinating to anyone who has reflected on its elusive but powerful influence. Smells and scents are the world's best social icebreakers. When people hear what I'm researching, they immediately start talking. First there are discussions about food and wine, but pretty quickly they move on to personal topics: about the smells that evoke memories of childhood, dead relatives, pregnancy, love and sex. When people talk about their sense of smell, they also tell us something about themselves, their feelings, thoughts and relationships. Thus, the sense of smell not only reflects the chemical world outside us, but also bears witness to our inner world, our feelings and thoughts.

In the following chapters, we will explore the little-known depths of scent and how it shapes us: Why can smells evoke such strong memories? How much can body odors control who you fall in love with? How sensitive is the human sense of smell? Can smells make us work more efficiently, feel better—and keep the brain vital? How can the smell of our own children fill us with such strong emotions? Why can a sudden bout of nausea cause the smell of a favorite dish to become intolerable instead? And what are the nasal superpowers that allow wine tasters to pick out the grape varieties, regions and production processes of an ordinary wine? These are some of the questions that will be addressed.

My twenty-year research journey into the world of smell has been full of surprises, and with each question answered, several new questions have arisen. It is a journey that has broadened my view of what the sense of smell is. My insights shaped the content of the book you have in front of you. Three messages will permeate the book. The first is that the human sense of smell is extraordinarily sensitive. We have an incredibly well-developed ability to sense smells in our environment. So I want to dispel the myth that humans have a poor sense of smell. Fortunately, many people agree with me, and psychologists and brain researchers are now busy reassessing both human sensitivity to smell and the importance of the sense of smell in our lives.

The second message is that smelling is an intellectual process that starts in the brain. When I started my own research, I viewed the sense of smell as a simple, passive system: An olfactory impression starts with airborne molecules that travel through the nose and get stuck in the olfactory mucosa of the nasal cavity. This sets off a biological chain reaction that leads to signals in the brain, where our *conscious olfactory* experiences occur as reactions. This is the traditional view of the sense of smell. However, my research has increasingly led me to adopt a different perspective: olfactory processes start not in the nose but in the brain, even *before the odor molecules reach their destination*. Olfactory processes are shaped by expectations and experiences, which in turn carry with them life experiences and cultural conventions that we are often unaware of.

I call it the *cognitive perspective*. *Cognition* comes from the Latin *cognoscere*, which means "to learn," "to know" or "to recognize." Cognitive scientists have long wondered how we gain knowledge of the world through our senses, our thinking and our brains. In this book, I want to use the cognitive perspective to understand the sense of smell. I believe that the human sense of smell is smarter than you think. It has its own kind of intelligence.

The third message is that we humans can make our lives richer and better when we turn our attention to the sense of smell and its capabilities. The sense of smell is not immutable, but can be improved and refined using insights from psychology and brain research. It's a perspective that offers some hope to those whose sense of smell is naturally weak, or who have suffered a loss of smell after a viral infection. What can they do to start recognizing the differences between grape varieties in wine, experience the smell of fresh rain during a walk in the forest or connect with the smell of a newborn baby?

This is the subject of the third part of the book. There I describe the ability of the sense of smell to recover and improve, and how my research group is working to find new methods to train the brain using smells. In other words, nasal gymnastics. The olfactory training program that I developed to prevent memory impairment has had a completely different application since 2020. When the coronavirus hit the world, the sense of smell suddenly came into focus. Suddenly, more and more people contacted me, telling me how their sense of smell and taste had changed. Some described it as all food tasting like cardboard. Some suffered intense distaste when their boyfriend suddenly started to smell rotten. One became depressed because he could no longer smell his children. So many people who in a short time were affected by an unknown, and invisible, disability. They asked if I could help them in their rehabilitation.

At first I thought no. I'm not a physician and I don't prescribe medication. According to the conventional wisdom about the ailments of our bodies, the only cure for loss of smell must be a pill, a spray or a surgical procedure. But medical technology does not yet have much to offer those suffering from changes in their sense of smell. I noticed that the emails I received had a common theme: the sufferers had not received any help from medical professionals. They needed something that the health care system could not give them. They lacked words for their experiences, they lacked recognition from friends and they missed the memories and emotions associated with smells that could no longer be brought to life. I realized that my experience and knowledge could be useful to them, and also to anyone who knows someone suffering from this loss. The emails from sufferers gave me the impetus to write this book. One of the main discoveries from my twenty years of experience as a smell researcher is that smelling is an ability that can be developed, improved and refined. Experiences are something we can talk about and share with others—if we find the right words. Therefore, there are also unexpected opportunities to cultivate the sense of smell—a journey that starts on the next page.

PART I

THE FORGOTTEN SENSE

CHAPTER 1

THE SUPERPOWERS OF THE SENSE OF SMELL

I F YOU HAD to sacrifice one of your senses, which one would you choose? Think about it for a moment. Consider your senses and what experiences you would lose without them. Think about the sound of your favorite music or your child's clucking laughter. Sunrises over rooftops, pine forests or the horizon. The wind from the sea in your hair. The smell of freshly brewed coffee or your lover's skin. No, this is not a choice made lightly.

When 250 British adults were asked which sense is the most important, only two, less than 1 percent, chose the sense of smell. Fewer than any other sense. In 2012, when a consulting firm asked seven thousand young adults what they would prefer to keep if they could choose from a list that included their sense of smell, passport, makeup and cell phone, half were willing to sacrifice their sense of smell to keep their cell phone. In another survey, Americans were asked what was more important: their sense of smell or their pinky toe? The result was a tie—participants thought the little toe was just as valuable as the sense of smell. This was the case—at least until the coronavirus pandemic broke out in earnest in 2020. Since then, around 300 million people have suffered from an impaired or distorted sense of smell after contracting the coronavirus. Most people now know someone who has been affected. Has this made us care about the sense of smell? Yes; a research study conducted in the US in 2021 by Rachel Herz and colleagues showed that the sense of smell may now be making a comeback. Only 15 percent of respondents chose to sacrifice their sense of smell for their pinky toe, and only 19 percent chose to sacrifice their sense of smell for their phone. Admittedly, sight and hearing were still considered far more important than smell. But perhaps we are finally rediscovering the importance of the devalued sense of smell. How did it end up at the bottom of our list of important sensory experiences? To understand this, we need to travel to nineteenth-century Paris.

On May 19, 1859, the Society of Anthropology of Paris held their inaugural meeting. Paul Broca took the podium, well aware of the spy sitting in the audience. He was familiar with the other audience members. Nineteen respected colleagues, well-dressed and mustachioed. Together with Broca, they formed a significant part of Paris's scientific elite. They all knew each other. But among them there was also an unknown man. The spy was sent, Broca knew, by the chief of police on the orders of the bishops of the French senate. The spy's mission? To ensure that the society did not fall into "atheism and materialism."

By this time, modern science had rocked the world. Charles Darwin, Broca's role model, had recently presented the seeds of the theory of evolution. In a presentation of his research, Darwin had argued that animal species changed through evolution, and this notion would lead to the controversial conclusion that humans were related to apes and other animals. *On the Origin of Species* would be a bombshell when it was released in the fall. Broca was a doctor and physiologist, and like Darwin, he was fascinated by animal breeding.

But the time was not quite right for his ideas. Broca stirred up controversy in the French biological society with a paper on how hares and rabbits could undoubtedly produce a common offspring. According to Broca, these hybrids showed that the boundaries between different animal species were not as strict as previously thought. The established view was that animal species were created by God, each with a specific purpose in creation. The paper was considered both embarrassing and blasphemous. Biologists were not ready for this kind of innovation.

After long negotiations, Broca had managed to get permission to form his own scientific society, and organize the first meeting of la Société d'Anthropologie de Paris, the world's first association to understand the innermost nature of man by studying his behavior and brain. The society was an initiative of freethinkers among Broca's colleagues, and they were not popular in all camps. According to the men of the Church, freely seeking the truth about humans risked giving rise to subversive ideas, such as that humans do not have free will. Paul Broca's own research laboratory had long been under surveillance. The authorities did not allow societies of more than twenty people, and all members of the society had been thoroughly vetted by the authorities.

Paul Broca studied the brains of many different human and animal species and understood better than anyone else at the time how they differed from each other. That is, he didn't understand much. Among his papers were detailed studies of the brains and craniums of everyone from the Italian writer Dante Alighieri to "the murderer Lemaire," whose particular characteristics and actions Broca thought he could understand from their brains. Political ideas were rampant in brain research at the time, and the presence of the Church often required researchers to package their findings creatively to avoid criticism. Broca's investigation of the differences between French and German brains made him popular with patriotic countrymen.

The French had wider heads and therefore better brain functions, Broca claimed. But a particularly big research problem was the fact that different animal species had such similar brains. It was common knowledge among experts that the brains of humans and apes were similar, and this fact risked leading to the uncomfortable conclusion that they were cast in the same mold, that they were related and that humans might lack a unique and free will—thus questioning an important dogma for the Church. It was important to avoid conflict with churchmen, and after the Paris meeting, Broca navigated the minefield. Likening humans to other animals could lead to problems. The solution came down to the sense of smell.

The human brain, Broca said, was characterized by large, powerful frontal lobes. People who suffered damage there lost the qualities we consider uniquely human: decision-making, social skills, language skills. Simpler animals did not have such impressive frontal lobes, nor did they exhibit anything resembling human intelligence. What the simpler animals had instead, Broca said, were *olfactory brains*. In the rat and mouse, the olfactory bulbs, the buds that receive the olfactory signals from the nose, were mighty bulges in front of the rest of the brain. But in humans, Broca argued, the sense of smell had receded to make room for the frontal lobes. He wrote: *This lobe, enlarged at the expense of others, occupied the cerebral hegemony; intellectual life is centered there; it is no longer the sense of smell that governs the animal, it is intelligence enlightened by all the senses.* Broca came to divide animals into two categories: the *osmatic* animals, which have the sense of smell as their main sense, and a few *anosmatic* animals, which, like humans, are not controlled by their sense of smell. The division was based on the opposition between the sense of smell, which Broca considered "bestial," and higher intelligence. It was intended to explain how humans differed from other animals: we humans are governed by reflection and thought and have suppressed the primitive sense of smell. The division was also colored by the racial theories of the time.

Some of Broca's anthropological colleagues had started mapping how smells had important cultural and religious meanings in other, non-European cultures. The importance of smell was interpreted as a clear sign that these peoples were more primitive than Europeans. Contrasting the primitive sense of smell with the intelligence and civilization of the Europeans was a compelling comparison, understandable and appealing to both the public and the most zealous controllers of the Church. The neuroscientist John McGann has argued that Broca's distinction between osmatic and anosmatic animals was so appealing that it came to shape the view of the human sense of smell for 150 years. So strong that even today young people often value their cell phones more than their sense of smell. The problem is that this comparison led us astray.

Broca's great discovery, that the frontal lobes of the brain contain much of what we consider to be uniquely human characteristics, also meant that another part, the olfactory brain, was overshadowed. This created an opposition between the higher intellectual abilities and the lower sense of smell. The British physician and anatomist Sir William Turner took over from Broca and made a new classification: animals had either a sensitive sense of smell, an insensitive sense of smell or no sense of smell at all. Turner placed all animals in one of these three categories, but he ignored the fact that there were no scientific studies yet on which animals had a better or worse sense of smell. It was a guessing game disguised as science. Turner based his classification on Broca's division into osmatic and anosmatic animals. But there was an insidious slippage from Broca's definition of "being *controlled* by the sense of smell" to Turner's definition of "having a *sensitive* sense of smell"—which are two completely different things.

We have a bad sense of smell because we are so smart! The notion became part of the legacy of Paul Broca. The irresistibly flattering narrative reappeared in many medical textbooks and treatises around

1900. The American anatomist C. J. Herrick went all out when he wrote that man's sense of smell was "greatly impaired, almost rudimentary" and that "the vastly greater equipment of most other mammals gives them powers far beyond our comprehension." Man's supposedly poor sense of smell even interested Sigmund Freud. For him, it was necessary for modern man to suppress the sense of smell, as it reminded us of our animal urges from when we were four-legged and, like dogs, sniffed each other instead of saying hello.

The sense of smell—primitive and instinctive. Language—reasoning and intellectual. Smell and language were presented as opposites. Even when comparative research results pointed to a well-developed human sense of smell, it was difficult to let go of what had now become dogma. Sir Victor Negus compared the olfactory mucosa of rabbits and humans—the area of the nasal cavity where smells are captured and transmitted to the brain—and found that it was actually larger in humans. This could have led to a reassessment. But strangely enough, Negus came to the opposite conclusion, that "in man the sense of smell is weak and not of great importance," and that studying the human sense of smell was a waste of time. Such was the power of theories that scientists even ignored results that defied their own prejudices. When scientists distanced themselves from the sense of smell, they also distanced themselves from our animal prehistory, and it became a way to emphasize the uniqueness of humans over other animals.

It was at another gathering, 156 years after Broca's Paris meeting, that the myth of man's poor sense of smell was punctured. It happened on Thursday, April 23, 2015, at a hotel in Bonita Springs, Florida, where the Association for Chemoreception Sciences was holding its annual meeting. A small buzz went through the audience when zoologist Matthias Laska from Linköping University changed the PowerPoint slide. The slide showed a funnel-like diagram, a summary of the research conducted so far on the sensitivity of humans and other animals to smell, and its meaning was immediately clear. Laska explained that many different animal species had been tested for their sensitivity to different odor molecules. Of these species, a total of twenty of them had been tested on the same odor molecules that had been tested on humans, making a comparison possible. Who was most sensitive? Sitting in the audience at the time, I read with amazement which species had been tested. Monkeys. Vampire bats. Dogs. And, of course, rats and mice, the most common laboratory animals, which have taught us so much about the functions of the body and brain. But what impressed the audience was how unambiguous the results were. Humans' sense of smell was clearly better than that of most other animals! The scientists in the audience had spent their working lives investigating the senses of smell and taste. Many had probably suspected that the human sense of smell was better than its reputation. But still, Laska's compilation was a revelation: humans have an excellent sense of smell.

Because of Laska's research, the old myth was blown out of the water. Humans were more sensitive than other animals to the vast majority of odor molecules. Human versus rat: 31 to 10. Human versus mouse: 36 to 35. Human versus vampire bat: 14 to 1. Human versus spider monkey: 58 to 23. The only animal that was clearly better than humans was the dog, which is more sensitive to 10 out of 15 odor molecules. Dogs are the olfactory kings of the animal kingdom, and later on you will learn about their special olfactory abilities. Perhaps the most surprising aspect of the comparison is that we humans actually perform so well in the smell tests that we can even give dogs a run for their money.

But perhaps you are wondering how it is possible to investigate the odor sensitivity of a rat. They can't tell us in words which smells they're most sensitive to. And how do we know that the rats really care about the researcher's odor task and aren't thinking about anything else? It is actually quite easy to test their odor sensitivity, but it takes some time. First, a rat must learn to associate an odor with food or drink. The researcher usually lets the animal go hungry or thirsty for a few hours before the experiment; this is not dangerous for the rat, but it increases its interest in exploring its surroundings. The rat is then released into a small chamber with a hole in the wall and a water bottle. When the rat sticks its nose into the hole, a smell machine puffs out an odor, and then the water bottle releases a drop of water. This does not happen every time. Sometimes only clean air comes out of the hole and the rat gets no water. Finally, after practicing for a while every day for a few days, the rat gets the hang of it: when the smell comes, I get a sip of water. Then the researchers see that the rat's behavior changes. When an odor is released, the rat rushes over to the water bottle, but when no odor occurs, the rat pulls its head back and tries its luck again. Once the rat learns the connection, the researcher can reduce the strength of the odor. When the smell is weak enough, the rat reacts as if there is no smell there. Then the researchers have found the rat's odor threshold. Similar tasks can be designed for any animal that can smell and learn—and that's most of them. Elephants, for example, have an excellent sense of smell and they put their trunks into different smell boxes to locate the smell. If they choose the right box, they get a carrot as a reward.

Those who described human beings as odor-impaired animals did not do so on the basis of careful scientific comparisons, but on the basis of their own cultural prejudices and beliefs. Observations of the brain became, as so often in human history, the "evidence" to justify what was perceived as a hierarchy in nature. But when we take a closer look at the human olfactory brain, we understand a little better how the early scientists could be so wrong. The human olfactory bulb looks undeniably modest where it is tucked under the huge frontal lobes. It is natural for scientists to compare the size of brain areas with each other to understand them better. The space occupied by different areas is described in proportion to the whole brain. This habit is due to the fact that brain areas often change together during evolution. And the size of the brain is largely related to the size of the body. Larger animals have larger brains. Therefore, scientists adjust for brain size when assessing different brain areas and how they differ among animal species.

Your olfactory bulbs represent only 0.01 percent (one ten-thousandth) of the weight of the entire brain; they occupy about 60 cubic millimeters of volume. In a mouse, the weight of the olfactory bulbs is as much as 2 percent of the brain and the bulbs are easy to identify. Using this comparative research method, the mouse's sense of smell appears to be much better than that of humans. But on reflection, it is not always a good idea to think this way. After all, the mouse olfactory bulbs are not larger than ours, but much smaller. It's because the rest of the human brain is so much bigger that our olfactory bulbs look so modest. However, a large brain is hardly detrimental to the sense of smell. The ability of the olfactory bulbs to process odors depends mainly on the capacity of the interconnected neurons located there. So how many neurons are inside the human and mouse olfactory bulbs? It turns out that a human olfactory bulb contains 10 million neurons, while a mouse's contains . . . 10 million neurons. The typical mouse has the same number of neurons in the olfactory bulb as does the typical human in theirs, and according to Laska's comparison, we have roughly the same sensitivity to smell. In fact, virtually all mammals have a similar number of neurons in the olfactory bulb, even though the total number of neurons as well as the size of the rest of the brain and body vary greatly among mammals. Scientists tend to describe the olfactory brain as an exception, because although the rest of the human brain has become larger through evolution, the olfactory brain has retained its basic design. It has not regressed, but neither has it grown in the same way as many other parts of the brain. The olfactory brain has been preserved. It connects us humans with mice and other mammals.

If we are not "anosmatic" animals as brain scientists used to claim, but rather have a sharper sense of smell than many other animals, we can suspect that it is our large brains that help us to smell. Shouldn't the sense of smell benefit from the high brain capacity that evolution has given us humans? That would be the opposite of what scientists have long claimed, that the olfactory brain has been overshadowed by our big brains. Instead of describing the functions of the brain as a tug-of-war between different areas, as some of the pioneers of brain research did, Laska's compilation helped to create a different picture, where the sense of smell could be enhanced by the enormous capacity of the rest of the human brain.

So how good is the human sense of smell? Well, good enough to detect the odorant butyl mercaptan diluted to a concentration of *0.3 parts per billion*. This is equivalent to three drops of the substance diluted in a volume of air equivalent to an Olympic swimming pool. Our sensitivity to this foul-smelling substance, commonly described as smelling like a rotten egg, has given it a special application. Different varieties of mercaptan odors are added to household gas, which is naturally odorless, to reveal if there is a leak in the kitchen. In the early 1900s, parts of Stockholm were evacuated because a bad smell was spreading in the city. In fact, a small painting company had poured some buckets of paint and other materials—containing mercaptan—into the Riddarfjärden bay.

The mercaptan smell has saved many lives by warning of gas leaks. Unfortunately, we also encounter these substances in our lungs and intestinal gases, making some body odors particularly repulsive to sensitive noses. In the next few chapters, we will delve more deeply into the psychology of disgusting body odors.

Because scientists have not prioritized the sense of smell, myths have arisen in the place of knowledge. One of the most common myths is that "humans can smell ten thousand odors." This may sound like a high number, but our senses of sight and hearing have a much higher capacity than that. Many people have heard this estimate of the sense of smell, but few know where it comes from. The origin of this myth—because it is a myth—is an almost century-old research report. The scientists who wrote the report believed that there were four basic smells, which could be combined in a variety of ways to create all kinds of different olfactory experiences. They also argued that the sensitivity to these four basic odors was high enough for a typical human to distinguish nine different strengths. This became the input for the researchers' calculation, which showed that people can distinguish between 9^4, or 6,561, odors. This figure was rounded up to ten thousand, and this figure lived on in newspaper articles about the sense of smell. As you might guess, the estimate is highly questionable. For starters, why did scientists think there were four basic smells? Perhaps because it was believed at the time that there were four basic tastes: salty, sweet, sour and bitter. The sense of sight also has a few basic elements. What we see is based on the wavelengths of light, which are translated in our retinas into three basic colors: red, yellow and blue. These can be mixed into all different shades of color, and we can tell the difference between about 5 million of them. However, recent research has shown that smells cannot be reduced to a few categories at all. This makes the sense of smell more elusive than taste and sight. For the sense of smell, the simple building blocks are missing; instead, it is believed that the nuances of smells vary along several different dimensions. The two main dimensions correspond to pleasantness (how good or disgusting something smells) and edibility (how edible or inedible something smells). These two dimensions explain part of how we perceive smells. But then, according to some scientists, we have another twenty-two dimensions that we cannot describe in words, nor do they correspond to any simple chemical principles. So the sense of smell seems to be much more complex than both sight and taste.

Exactly how many smells a typical human can distinguish is still a mystery. But it is probably a very high number, much higher than ten thousand. A few years ago, researchers at the Rockefeller University tried, for the first time using advanced methods, to examine the ten thousand figure and find the true upper limit of human olfactory capacity. Selecting thousands of odor molecules and presenting them to participants would have been impossible for practical reasons. Instead, the researchers created a smaller set of 128 odorant molecules, diluted them to the same odor strength and then combined the molecules in mixtures of 10, 20 and 30 molecules each. An important aspect of the experiment was that the mixtures were partly composed of the same molecules, and the overlap made the mixtures harder to distinguish. The participants' task was just that, to try to distinguish odor mixtures that smelled similar. The results showed that if more than half of the molecules were the same, the mixtures became difficult to distinguish from each other. The researchers then performed an advanced mathematical calculation that led to an estimate of how many smells a typical human can distinguish: at least 1 trillion, *one thousand billion smells*! The figure is staggering. If true, smell is probably the most sensitive sense in humans. But the study has also been criticized. It relies on a very sophisticated mathematical model, and does not provide direct evidence that humans can really distinguish so many smells. Critics have shown that other calculations can result in much lower estimates. The mystery remains. But again, the figure is probably much higher than ten thousand. We can probably distinguish at least as many smells as we can colors— millions of them.

The human sense of smell can be used for tasks that we didn't ever think we could master. Take tracking, for example, where dogs sniff their way around, moving their heads from side to side and head-ing toward their prey. Neuroscientist Noam Sobel and his colleagues wondered if humans could perform a similar feat. They had students at the University of California at Berkeley smell their way around one of the university's lawns. For the experiment, the researchers laid a 10-meter-long trail of chocolate in the grass. The participants were astonishingly adept—21 of the 32 participants were able to follow the trail all the way to the finish line. Some of them then had to practice the task every day, and they became better and better "tracking dogs" the more they trained.

But human sniffer dogs are not the strangest example of human odor sensitivity. That prize goes to researchers at the Swedish University of Agricultural Sciences (SLU) in Alnarp, near Lund, Sweden. The researchers studied fruit flies, the small creatures that in summer smell their way to our fruit bowls, food scraps and wineglasses, which are excellent sources of nutrition for them. The females, the SLU researchers knew, emit pheromones, mating signals that attract the males. The researchers discovered that *they themselves* could smell the difference between females and males. But instead of letting this observation stay within the group, they decided to take a different approach. They convened a group of wine tasters to create scientific evidence that we can smell the pheromone odor of fruit flies. The researchers subjected the wine tasters to three different tasks to identify the fruit fly's smell: (1) in an empty wineglass where a fruit fly was held captive for five minutes, (2) in a wineglass filled with water where the fruit fly was bathed for five minutes and (3) in a wineglass filled with German white wine where the fruit fly was bathed for five minutes. Could the wine tasters smell the glasses where the fruit fly had been? The results confirmed the researchers' observations: the olfactory panel could easily recognize this pheromone substance in all three tasks. One nanogram—one-billionth of a gram—of the pheromone was enough to affect the smell of wine. The researchers concluded that the human nose is so sensitive that a single female fruit fly in a glass of wine can affect how we perceive the smell of wine! The discovery that humans can thus determine the sex of flies by their smell made them worthy winners of the Ig Nobel Prize, which rewards "research that first makes you laugh, then think." The Lund researchers' crazy experiments made them worthy winners, because isn't it amazing that our human noses are so sensitive that we can learn to recognize the fruit fly's chemical mating signals?

Besides historical prejudices, there is another reason why we underestimate the sense of smell. We refer to smells as *tastes*. Think about the last time you had a really bad cold, with a stuffy nose. Didn't your morning coffee taste like fizzy hot water? But how can the *taste* be changed by something happening in the *nose*? The taste buds are located on the tongue and are not usually affected by colds. This example provides an important clue as to why the sense of smell has been so underestimated for so long. The taste of coffee does not change with the cold. With our tongues we can experience sensations like sweet, salty, sour, bitter and umami. Coffee does not have a strong taste, just a bit bitter, and perhaps a bit sweet for those who drink it with milk or sugar. But coffee has a very strong *smell*. It reaches us not only through the nose but also through the throat when we drink. This is because the sense of smell has a "back door" to the brain through the throat. So when we eat and drink, we smell both through our nose and our throat. In humans and other primates, odors are released in the mouth when we eat, stimulating the olfactory receptors in the nasal cavity. How does this work? Think of the mouth as an air pump, pumping air through the throat, up into the nasal cavity and out through the nose as you chew. Chewing has released the food's odor and flavor compounds, and these follow the air up into the nasal cavity—but only if you eat with your mouth closed, otherwise the odor molecules go out the same way they came in. The same applies when coffee or any other drink flows through the mouth and is swallowed. The smells that enter the nasal cavity via the throat are important to what we call "flavors." We tend to think they are tastes being perceived in the mouth, when in fact it is the sense of smell working in disguise. When patients go to the doctor complaining that the taste of something has disappeared, the doctor understands that it is actually the smell. Anyone who doubts that the "taste" of strawberries, coffee or cinnamon buns is really more about the smell from the throat can do a little experiment: Cover your nostrils, then put any food or drink in your mouth and eat or drink. Try to concentrate on the unique "taste." You will notice that it has lost its nuances. To demonstrate this phenomenon, olfactory scientist Bill Hansson has his students taste ketchup and mustard and say which is which. This may seem like a simple task, but with your nose pinched it is almost impossible. Ketchup and mustard usually have similar sweetness, acidity and saltiness—only the smells are different. Then let go of your nose and start breathing through it. A world of nuances is revealed as the airflow through the nose opens up.

It is a unique feature of the sense of smell that the same receptors can be stimulated by two very different methods. And the two methods have completely different functions. When we sniff—scientists call it orthonasal smelling—we smell things outside us, and the source of the smell can sometimes be far away. Through sniffing, we can detect a forest fire far away in the distance, or that the Indian restaurant around the corner is open. When we sniff, we connect the smell with what we see and hear. The aim is to identify what we are smelling and to understand where the smell is coming from. The smells from the throat—what scientists call retronasal smells—come later, when we have already decided what is in front of us and that we want to put it in our mouth. Neuroscientist Gordon Shepherd said that the retronasal smell is a human specialty. Unlike dogs', human olfactory receptors are located quite close to the pharynx, where retronasal smells originate. This allows us to appreciate a steak served medium rare, where the odor molecules from the surface of the meat are complemented by other odor molecules produced when we chew through its less-cooked center. However, the dog's olfactory system is focused on sniffing and the orthonasal smell, which may be one reason why it doesn't matter to a dog whether the meat is medium rare or well done. Thus, our sense of smell is not solely adapted to sniff our way through life, as dogs do. Instead, our sense of smell is specialized to also sense the retronasal smells and weave them together with tastes, sensations and the internal states of our bodies—hunger, thirst and pleasure.

So humans have a powerful sense of smell. But odors are so much more than what we feel when we sniff. In later chapters of the book, I will show that the experience of smell is not just a reaction to the molecules that hit the olfactory mucosa—it is an active, creative process in the brain that affects us, often quite unconsciously. For those who want to become aware of the importance of the sense of smell, the first step is to learn from history. What cultural significance has the sense of smell had throughout history and how did we forget the smells around us over the years? The cultural history of the sense of smell begins in antiquity, where written sources attest to the many and rich meanings of smells. The olfactory trail from antiquity leads us into the olfactory world of the Middle Ages, where holiness, sin and disease had their specific smells, and into the modern age, where Broca and his colleagues worked and where the eye, rather than the nose, became dominant.

CHAPTER 2

CULTURAL CHEMICALS

SMELL THE BOOK you are holding in your hand. Close your eyes and sniff deeply. Experience the smells of paper, glue and printing ink. They tell you something about how this book was made, about the raw materials, the manufacturing process and the transportation. But it is a story that is difficult to decipher, communicated in a language that few know. Despite our sensitive noses, we often have difficulty understanding what we are smelling. We live, for better or worse, in the age of the visual sense. For those who can read books, text is an unparalleled source of knowledge. But this was not always the case. General literacy is a fairly recent development. In many Western countries, it evolved in the eighteenth century, motivated mainly by religious desire, and was later reinforced by the public schools of the nineteenth century. Literacy meant that mass-produced texts such as the Bible, rather than orally transmitted stories, gradually took over the role of our civilization's memory bank. Historians say that as we began to read, the sense of smell gradually lost its importance. Nowadays we rely mostly on our eyes and ears. The television and the personal computer became the most important communication tools of the twentieth century, along with the radio and the telephone. Our own century is so far dominated by mobile phones, with visually stimulating apps such as Instagram and TikTok enhancing and distorting our bodies and faces and disseminating them to a global audience. We are so deeply embedded in the age of sight that it is hard to see how it could be otherwise. The thought of a world where many of the important events in our lives revolve around smells seems alien. How did we get here? Is it possible to imagine that the sense of smell will return to prominence?

The devaluation of the sense of smell began long before nineteenth-century scientists put the final nail in the coffin of smell. Ancient philosophers had a clear ranking of the senses, based on what was considered more or less important. Smells were often discussed in passing, with idle interest. Aristotle wrote that our sense of smell is inferior to that of all other animals, and also inferior to all our other senses. For Aristotle, it seemed natural to claim that women had a better sense of smell than men because they were closer to animals and nature. Men, who according to Aristotle were better suited to intellectual activity, had sharper vision. The sense of sight was linked to abstract thinking, which was considered a male activity. The devaluation of the sense of smell was linked to the view of women. Philosophers had a negative attitude toward both. But the world that ancient people lived in was full of smells, odors and stenches.

If you were told that the Savior of humanity had been born, what gift would you bring? Two thousand years ago, a group of men were given a reason to ponder this stress-inducing question—the world's first Christmas gift anxiety. The story is well known. A star rose in the sky, and if we are to believe the Gospel of Matthew, some Eastern stargazers interpreted it as a sign that the Savior had been born. The men traveled to Bethlehem, where they found Joseph and Mary with the baby. There they left three legendary gifts for the newborn Savior: gold, frankincense and myrrh. It is a choice of gifts that is difficult to understand today. After all, everyone can understand the value of gold—it is still a hard currency. But incense and myrrh? Two of the three gifts from the wise men were *scents*.

For ancient people, smells were loaded with important cultural meanings. Pleasant smells were used to appease gods in religious rituals and in royal parades. Rich people's homes were beautified by being drenched in fragrance. Scented water was poured over the stands of the amphitheaters and the Colosseum. It helped to create strong and pleasurable experiences for the spectators—and perhaps to mask the less pleasant smells of the bloody spectacles. The scents used in ancient Athens and Rome came mainly from flowers such as crocus, hyacinth, rose and violet. Cinnamon and other spices were imported from the East, as was resin and other scented plant extracts, often at great expense. Myrrh is one of the fragrant resins that were used in health-giving oils. Incense was used in religious rituals and certainly evoked strong emotions as believers approached their gods in spiritual experiences. The ancient olfactory culture was largely inspired by the cities of the Middle East, where spices, incense and perfume oils competed with the natural smells of latrines, livestock and people engaged in hard physical labor. Ancient smells were literally impregnated into citizens' clothing—the word *perfume* comes from this, the Latin roots of the word meaning "smoke through."

The ancient poet Lucretius wrote that the human soul relates to the body as smell relates to a stick of incense. It is a beautiful thought, that the soul somehow emanates from our physical body like the body's odors, and that it gives an invisible energy, a feeling, to those close to us.

One can easily understand why smells were often used in religious ceremonies. The ancient gods apparently had an insatiable appetite for smells. Animals were sacrificed and burnt so that the strong smell of burning flesh could reach the heavens, where the gods resided. However, for vegetarians of the time—such as the philosophical brotherhood of the Pythagoreans—plant-based smells from flowers and herbs were sufficient. Smells could also be a form of transportation. They could help the soul leave the body after death and guide it to heaven. This is why perfumes, myrrh and other scents were used in wreaths at funerals and cremations. So myrrh was a great gift at this time. Expensive and exclusive, but also with a spiritual and deep cultural meaning that connected ordinary people with the gods, legends and myths.

Ancient philosophers sometimes reasoned about the essential nature of smells, what they really were. In their thoughts we find the seeds of the cognitive approach that I use to understand the sense of smell. According to the ancient worldview, the world consisted of only four elements: air, water, earth and fire. Philosophers struggled to fit smells into this rather limited conceptual world. Democritus believed that different smells could mimic the different elements, which could also explain their different emotional charges; pleasant smells consisted of round particles, just as he imagined water particles to be round, while unpleasant smells consisted of pointed, pyramid-shaped particles, just like fire. Plato's description of odors was quite different. He did not describe smells in terms of a particular form but as a process, something that happens when matter changes from one state to another, for example through decay, evaporation or cooking. Thus, for Plato, smells were not a material substance but more like a coincidence, and this made smells less interesting to him because he wanted to understand the eternal principles of existence. However, the thinker who had the best insights into the sense of smell at this time was Aristotle, even though he dismissed the human sense of smell as primitive and underdeveloped. According to him, there were certain smells whose emotional effects depended on the state of the body; for example, the smell of roasting meat could lead to different reactions depending on whether we are hungry or full. Other smells, however, were related to the material properties of the smells and not to the state of the human body. For example, the smell of decay was harmful regardless of who smelled it. This was a small but important shift—according to Aristotle, the meaning of smells was not entirely material, but arose from an interaction between the odorants and the human body. Modern science would prove Aristotle right. The human sense of smell varies greatly from person to person, and is influenced by our hunger and thirst and by past memories and associations. But, as noted by philosopher Ann-Sophie Barwich, the philosophy of smell was not the main interest of the ancient philosophers. They did not pay much attention to the sense of smell. Yet, almost in passing, they planted the seeds of our modern cognitive understanding of how the sense of smell works.

Perfumes were not appreciated by everyone. They were often associated with prostitution, because the prostitutes of the time were described as being drenched in perfume. The disreputable women were seen as trying to make themselves appear better than they really were. Perfumes came to be associated with a kind of falsehood—a perfumed person was someone who was trying to hide their true nature. Early Christians were therefore often opposed to scents and perfumes. They were considered to arouse animal instincts and attract people to sin. Just as today, in ancient times there must have been "perfume allergy sufferers," i.e., people who are hypersensitive to certain fragrances and cannot tolerate exposure to them. One such person is Cynculus, a character in a drama by Athenaeus, who, at a dinner party, frowns upon the obligatory perfuming of the diners: "Can't someone bring me a sponge and wipe my face, which has been contaminated by this dirt!"

In antiquity, then, we see that ideas about the sense of smell began to develop, although these ideas were limited by the prejudices of the time regarding men and women, and the fact that very little was known about what smells actually were. However, these ideas led to a simple "psychology of smell," whereby smells were characterized by their emotional properties, and these were determined both by the material properties of the smells and by the person smelling them.

Feeling weak, tired and depressed? Take a bunch of juniper rice and a bunch of sage and burn it. Breathe in the smoke. It will awaken your brain, strengthen your muscles, lift your spirits and prolong your life! Such exhortations could be found in medieval medical texts. According to medieval medicine, odors were physical substances that literally flew in through the nose and into the brain. Smells inhaled through the nose therefore had an immediate impact on the brain, with direct consequences for our health and state of mind. The theory is called *miasma,* and various such beliefs have been found in different parts of the world and in different times. Pleasant smells were believed to be good for the brain and body, and many doctors prescribed scent cures for their patients to treat illnesses. On the other hand, bad smells— from stagnant water, rotting food, abscesses—were considered bad for the brain and body. So when the plague ravaged Europe in the seventeenth century, it was the smells associated with disease and decay that were thought to be dangerous. But there were also smells that were thought to protect against and even cure the disease. Doctors carried an "odor bag" under their nose when they visited their patients. The theory was influential for a long time, and so-called "beak doctors" were sometimes seen during the plague. They wore a face mask with a large bird beak filled with lavender and other dried herbs and flowers. Breathing through the beak was thought to clear the air of disease—since viruses and bacteria were unknown at the time, there was no distinction between disease and its smell. In this way, it was believed that the beak would keep the doctor healthy. Some believed that all strong smells, even the most disgusting, offered potential protection against disease, and so some health-conscious individuals could be seen in the mornings bending over a latrine to inhale as many strong smells as possible in order to be protected for the rest of the day.

Body odor was a kind of medical diagnostic tool in the Middle Ages. The scientific theory of body odor was based on the ancient physician Galen's theory of the four bodily fluids: blood, mucus, yellow bile and black bile. The four fluids were in turn linked to the ancient concept of the four elements: air to blood; water to phlegm; fire to yellow bile; and earth to black bile. The balance of the four bodily fluids affected how people smelled, and what smells attracted them. Diseases disrupted the balance of body fluids and therefore created abnormal smells. For example, a person with leprosy was considered to have too much black bile and therefore emitted a bad smell. We now know that the leprosy—and the smell associated with it—originated from a bacterial infection. But the miasma theory of the sense of smell made the nose an important instrument for medieval doctors. Until the eighteenth century, it was common for doctors to smell their patients to understand what diseases they were suffering from. The doctors' sensory experience was more important than today. The disease diphtheria smelled sweet, scurvy smelled pungent, typhus smelled like freshly baked rye bread, and scrofula, a type of tuberculosis, smelled like stale beer. The historian of ideas Karin Johannisson writes in her book *Tecknen (Signs)* about how doctors of the time had a more personal relationship with their patients; they squeezed, observed, smelled and tasted. It was natural for medieval doctors to smell their patients and taste their urine in order to make a diagnosis. They prescribed certain smells as medicine to their patients and advised them to avoid others.

A common thread throughout history has been the attribution of health-promoting properties to odorous substances. Bad smells reveal disease and decay, while pleasant smells characterize vigor and health. In a way, it is natural to think this way, as diseases are often accompanied by specific smells. Think of the smell of decay in someone with infected wounds or bad teeth, or in some elderly persons, where the body's increasing inflammation characterizes the body's smell. But people in ancient times did not have the same understanding of causality as we do today. Therefore, they prescribed ineffective scent treatments. In ancient times, it was considered appropriate to bathe burns with a mixture of wine and myrrh because of its smell. Yes, the smell of burnt flesh and scarring was certainly mitigated by this mixture, but its healing power is questionable. The smell of apples was considered an effective treatment for gout. By treating the bad smell, people thought to treat the disease itself.

Body odors were not only important signals of human health, but were also related to a person's appearance and temperament. Smells were even thought to reveal people's moral character. The clearest example of how smells reflect the characteristics of the soul is the medieval stories of the lives of saints. A clear theme in the religious texts of the time is that sweet, flowery scents are associated with holiness and the smell of decay is associated with evil. Strong smells of fire and sulfur were associated with tortured sinners in hell, while pleasant smells delighted the holy souls in heaven. Medieval plays and stories often tell of devils and other evil characters releasing foul-smelling farts, but around the Virgin Mary and the saints the scent of delicious spices spread. Thus, a person's smell was an important sign of holiness, even after death. According to the poets of the time, the bodies of saints did not decompose and smell rotten like ordinary corpses. Instead, they often had a "smell of holiness"—usually of flowers. And since body odors could reveal spiritual qualities, the sense of smell also played an important role in evaluating a person's character.

However, smells could also be dangerous and misleading. Attractive smells from herbs, spices or people could seduce others; remember the prostitutes with their strong perfumes. Medieval priests often warned people not to "sin with their noses." To counteract this, the priests thought that old meat and manure could be used to avoid temptation. There are stories of monks keeping rotting corpses in their rooms for this purpose.

It was not only good smells that led people to sin. Bad smells could also be misleading. For example, in Christian literature it was considered a sin to avoid sick people because of their smelly wounds or bad breath. Therefore, theologians urged their audience not only to smell with the physical nose itself, but also to use the "nose of the soul." This meant not only judging smells based on immediate emotional experiences, but also trying to find moral meaning in smells. The body may want to avoid disgusting smells, but those close to God could override these feelings and do the right thing. Saint Catherine of Siena was a fourteenth-century Italian nun who later became revered as a saint in the Roman Catholic Church. She was thought to be able to smell with the nose of the soul. Her biography tells the story of how she cares for a nun who has cancer. The nun smells so bad that all the other sisters are driven away. But Catherine stays—she doesn't even cover her nose. The smell is so strong that Catherine feels nauseous. But she understands that this is just a bodily odor that the devil is using to try to get her to abandon her duty. So she presses her face down on the area of the tumor until she no longer feels nauseous. In a medieval legend, we meet a young woman who does not want to get married. She stuffs rotten chicken breasts under her armpits to discourage her suitors. Unsurprisingly, the men think she is sick and therefore not a suitable candidate for marriage. But one of the men can smell her true odor through the stench of the bad chicken. They get married and live happily ever after. True love could overcome everything—even a foul odor.

Nowadays we often talk about being able to *highlight* or *visualize* a problem and *see* a solution. We gain *insights*. The language reveals how our different sensory experiences are thought to have different properties. The sense of sight seems to be a particularly intelligent sense, if we are to rely on everyday language. However, the senses of smell and taste are often "invisible" in such contexts. They are mostly used to describe emotions and values. When we say that a behavior *leaves a bad taste in the mouth*, the sense of taste is present to show our negative feelings. However, there is reason to believe that in the Middle Ages the sense of smell was also a respected source of knowledge. To some extent it was a matter of life circumstances; a person living in Europe in the thirteenth century had to rely on the world of knowledge offered by the immediate environment and their own sensory experiences. No printing presses could spread knowledge far and wide. Higher education and abstract knowledge were luxuries available only to a small elite. Few people could read. Long-distance travel was rare, and the dominant agricultural existence was a world of practical knowledge, rich in the smells of the soil, crops and animals. It is easy to imagine that smell and touch played a greater role in such a culture than in our own. The importance of the knowledge of the senses was reflected in the writings of scholars, sometimes in unexpected ways. For example, in the sixth century, the bishop and writer Isidore of Seville wrote a book on the origin of words, linking the Latin words *noscere,* "to know," and *naris,* "nostril." So, according to Isidore, knowing something meant being able to "use your nose." He called ignorant people *ignarus*—those without nostrils! (Isidore's interpretation has not been supported by modern linguistic research.) Today, the sense of smell is associated with intuition rather than reliable knowledge. The English language uses the expression *smells fishy* when there is a suspicion that something is not right. And those who *sniff out* something have found the answer through concentrated effort, not logic and reasoning. But even in the fifteenth century, a full eight hundred years after Isidore, angry, impolite, ridiculous or ignorant people could be described as "those without nostrils." Much later, in the late nineteenth century, the German philosopher Friedrich Nietzsche would claim to be able to sniff out truths and lies, and said "my genius lies in my nostrils."

This was a tribute to intuitive thinking and our instincts, which Nietzsche valued more than philosophical systems.

In a world of thought where smells played such an important role, it was perhaps only natural that the appearance of our noses was also thought to carry different messages about our inner qualities. Religious texts from the Middle Ages contain many references to noses, and capable priests were thought to be able to distinguish good from evil with their sense of smell. Today, these ideas seem alien. But our noses have instead been attributed with other important qualities. Jayakar Nayak, a surgeon and professor at Stanford University, has called the nose "the silent warrior: the gatekeeper of our bodies, pharmacist to our minds, and weather vane to our emotions." This is a beautiful and accurate description. The nose plays an important role in warming, humidifying and purifying the incoming air. The nasal hairs trap larger particles and prevent us from pulling them down into the lungs. Along with the air that travels into the nose, bacteria and viruses are also carried along, and we don't want them to reach the lungs. Therefore, inside the nasal cavity are the nasal turbinates. They are bulges lined with mucus that collect all the debris and send it down the throat toward the gastrointestinal tract, where it can be neutralized and disposed of at the next toilet visit. The air-warming and humidifying function of the nose has been particularly important in cold, dry climates. This is probably why the appearance of noses differs so notably among different ethnic groups. People of European origin are distinguished by their particularly long noses, which can warm the cold air they breathe. In the warm climates of Africa and Southeast Asia, noses are often smaller and flatter. Just as our skin has different colors depending on the climate our ancestors lived in, scientists believe that the nose has also adapted to the climate.

But our noses have also been the focus of harmful and unscientific ideas throughout history. The notion that the appearance of the nose reflects our personality was a theory that grew out of medieval ideas about the sense of smell. Paracelsus, the physician and natural philosopher, argued in the sixteenth century that the shape of the nose could reveal evil, falsity and mendacity. The eighteenth century saw the development of physiognomy, a theory that claimed it was possible to read personal traits in the face. The rise of physiognomy coincided with the migration of people from the countryside to the growing cities where the new industrial jobs were located. In the city streets, people from different backgrounds mixed, and poverty and crime were rampant. Who could be trusted? Could the shape of the nose reveal the impostor? These ideas were taken very seriously for centuries. As for the nose and nostrils, much was written about their size, shape and straightness—things that were thought to reveal a person's intelligence, emotions, combativeness or sexual drive. Although there is no scientific evidence that the appearance of the nose is related to character, such books were published as late as the 1980s. *Den praktiske "människokännaren,"* published in 1951 and reprinted in 1982, states: "A long, sharp nose combined with a flat, backward-tilted forehead and a similar chin: small-mindedness, egoism, coldness." And in *Secrets of the Face,* published in 1984, the author, Lailan Young, states that "an extremely downturned nose means that the owner is unreliable as a friend." These books did not have particularly high scientific ambitions, but it is still astonishing that such fabrications could be published so recently. Let's hope that the pseudoscience of physiognomy remains in the historical archive of bad ideas.

The sense of smell was relatively prominent in the medieval world of ideas, but this changed. The sense of sight became increasingly important with the scientific revolution in the seventeenth century. Technological innovations in glassmaking allowed Galileo Galilei to construct telescopes with up to twenty times magnification. These new telescopes led to a whole new scientific worldview. This new way of thinking was largely about transferring knowledge of space, which was now clearly visible in the telescopes, to life on Earth.

Previously, philosophers had made a sharp distinction between space, where celestial bodies were thought to move in perfect circular motions and could be described mathematically, and the Earth, where life was characterized by the chaotic interaction of the four elements, earth, air, fire and water, which stimulated all the senses but could not be measured. With the scientific revolution in the seventeenth century, this division broke down. Even life on Earth now began to be described using mathematics and geometry. The four elements were abandoned and replaced by measurable movements and forces. How fast do different objects fall? How much does air weigh? How fast do sound and light move? All this required accurate measuring instruments—which were read visually.

Modern developments thus led to a downgrading of the cultural importance of the sense of smell, and paradoxically, a contributing factor was the increasing demand for new and exciting smells. The spice trade between Europe and Asia had grown into a huge industry in ancient and medieval times. The demand for spices, but also herbs and resinous oils for cooking, religious and medicinal purposes, drove long-distance trade between Europe and the Far East. The need to find a more efficient route to the fragrant spices of India led Christopher Columbus to lead an expedition westward—the one that led him to America in 1492. The "new" continent, and its wildlife, unknown to Europeans, became a catalyst for modernization and the scientific revolution. It is ironic that the quest for better and cheaper smells also led to the neglect of the sense of smell.

Visual impressions came to increasingly dominate our world in the seventeenth century, and the new sciences contributed to the shift. Visual artists turned to mathematicians, who taught them how to apply geometric and optical principles to give paintings a perceived depth and more realistic appearance. Cartographers used the same geometric principles, thus developing methods to describe the appearance of the Earth's surface in a coordinate system. Good maps gave trading companies an edge over their competitors in the global world that opened up to Europeans as the New World was discovered. And when conflicts arose, Europeans had another ace up their sleeve: the geometric principles that had revolutionized painting and mapmaking had also made ballistics—the science of the trajectory of cannonballs through the air—a burgeoning field of research. All these mathematical innovations could be accurately illustrated in new books that, since the introduction of Johannes Gutenberg's printing press in the mid-fifteenth century, could be produced at an ever-increasing rate. This reinforced the idea that abstract thinking was linked to the sense of sight.

Medieval thought had engaged all five traditional senses, not least smell, but modern science focused mainly on sight. The decline of smell was not only scientific, but also political and economic. In France, the ruling class, the aristocracy, were known perfume enthusiasts, and the court at Versailles occasionally ordered everyone in the palace to change their perfume every day. But their habits were increasingly associated with corruption and oppression—a return to the old idea that perfume was a form of deception, a way of making oneself appear better than one really was. Odorless environments were also increasingly seen as a hygienic ideal. European cities developed sewage systems, and access to clean water made it possible to bathe and wash oneself and one's clothes at home, at least occasionally. The stench of cities decreased in the nineteenth century. There was less reason to drench oneself in smelly oils, smoke or sprays. The urban middle class had no contact with agriculture and livestock. Cleanliness was the new trend. Smells and odors belonged to the old, dirty world.

Medical science had also started to move away from theories about smells spreading disease. With the mass production of books, medical students could now acquire knowledge that way instead of, as before, studying by sitting next to a patient in bed and feeling, listening and smelling. It became clear that diseases were not caused by smells at all. Louis Pasteur showed in his experiments that bacteria were often the cause of illness. Odorants moved from the pharmacist's medical shelves to dedicated cosmetics stores. The sense of smell lost status on all fronts. It was in this new era that Paul Broca, whom we met in the opening chapter, formulated his theory that humans were unaffected by their sense of smell. It was a timely idea.

With the scientific revolution that began in the seventeenth century, humanity entered the "age of sight." At least, a first step was taken. Because the dominant role of the eyes in our culture has only continued since then. We now live among screens. Images and videos are becoming our main sources of information. However, smells cannot be so easily captured, re-created and disseminated with digital technology. There is no streaming service for smells. Today, a doctor's olfactory diagnosis is not taught in textbooks. And the miasma theory, that it was bad smells that caused plague and other serious diseases, was superseded, with good reason, by new biological knowledge. But in retrospect, we can still see that the miasma theory had some positive consequences, leading to important sanitation reforms. Improved sewage systems were motivated by the idea that the very smells of the stinking urban environment were harmful.

In other cultures, we encounter worlds of smell that are different from the Western one. Smells can support entire worldviews, and such olfactory worlds constitute what some anthropologists call *osmologies*. For the Indigenous people of the Andaman Islands, an archipelago in the Indian Ocean, smell is a way of understanding both time and space. They have constructed a calendar based on how nature's smells change between seasons. The plants that bloom give each season a particular energy through their smells. It is the energy of smells that keeps the world going and makes the seasons change. By eating seasonal fruits and plants, residents can get the necessary energy to maintain their health and vitality.

In the Andaman Islands, as in ancient Greece, smell is also a way to communicate with the spirit world. The people are very aware of their own smells, as calling the spirits at the wrong time can be dangerous. The spirits move across the Andaman Islands and they can smell where people live, so villagers try to minimize unnecessary smells. In religious rituals, the opposite is true, as smells are amplified to call the spirits' attention. Different smells for different places.

Those who seek examples of cultures where the sense of smell is central can find them in nonindustrialized societies and in our own history. Our culture, on the other hand, is now considered to be based on reason and enlightenment, framed by written words and images. How could a modern, high-tech world accommodate smells and their rich cultural meanings? And what would that mean? Anthropologist Constance Classen argues that valuing the sense of smell has a radical, almost subversive potential. According to Classen, the sense of smell defies the social order because it reminds us that there is another side to our lives where desire, pleasure and disgust take center stage. Valuing the sense of smell thus seems like an exciting, but perhaps also somewhat threatening, prospect.

The marginalized sense of smell has received an unexpected boost into the spotlight in recent years. The coronavirus pandemic deprived millions of people of their olfactory experiences and showed how important the sense of smell is to our lives. Smells are important sources of pleasure and will probably always be important to us. But the most crucial smells of the future will probably not be sniffed by ordinary flesh-and-blood noses. "Electronic noses" technology has made great strides in recent years. These are machines that can perform chemical analysis of airborne molecules. The e-noses were developed to mimic the capabilities of a human nose, with chemical receptors and a "brain"—the software program responsible for signal processing and recognition of key odorants. These machines proved early on to be able to detect lung cancer in patients' exhaled air. More recently, the e-noses were able to detect signs of COVID-19 infection with high accuracy. In addition to exhaled air, this technology has been used to "smell" pee and poop, and detect rectal, prostate and bladder cancer in its vapors. In the future, our home electronics will not only be able to image and listen to us, but perhaps even smell us. Small, portable computer noses could be connected to our mobile phones to analyze the chemistry of our breath. In the future, when you travel abroad, you may be allowed to breathe on your phone, which will quickly examine your breath and assess whether you have a cold or other viral disease that could be contagious. This "odor monitoring" may seem either promising or threatening, depending on who you ask.

But it is not only in medical diagnostics that e-noses can play an important role. E-noses are already used in modern factories to detect bacteria or mold in food and drinks. A Spanish research team recently developed a flying e-nose. It is mounted on a drone that flies around and measures air quality. This can give a better picture of the worst sources of emissions at wastewater treatment plants—which of the chimneys is the most polluting? But you could also let the flying nose freely explore its surroundings and detect unknown emissions and leaks. Flying noses protecting us from harmful emissions—a dizzying idea.

And let us return to our beloved screens, the very symbol of the hypnotic power of the visual sense. Will tomorrow's screen time bring us smells and tastes as well as visual stimulation? Yes, if you believe Japanese scientist Homei Miyashita. He has invented a screen that delivers taste and smell sensations. The device, called TTTV (Taste The TV), consists of an advanced smell and taste mixer that sprays its contents onto a film that is then rolled down a screen, ready for the user to lick. The screen simultaneously displays moving images of the food. A cheap and low-calorie alternative to going to a luxury restaurant? Well, Miyashita's invention is unlikely to compete with real restaurant experiences. For starters, it's very difficult to re-create the smell of well-cooked food with a few chemical ingredients. But perhaps TTTV could be used in online wine tasting and cooking classes? Another, more portable smell technology has been developed by technology company Sony. Aromastic is a small smell machine, designed in Japan, which is supposed to offer smells for various "aromatherapy" purposes—invigorating, relaxing, and so on. The idea is for the customer to carry Aromastic in their pocket and sniff the pleasant smells when needed, on the subway, in front of the computer or during a meeting. Aromastic originally cost about USD 90, and despite price reductions, it has not become a big seller for Sony. When I met the Aromastic team at Sony's headquarters in Tokyo, I gave them my best advice on how to make the product a success. For me, the weakness of Aromastic is that it "only" creates smells. Psychological research shows that for our smells to be transformed into rich experiences, they need to be used in the right context. The best olfactory experiences are created when the sense of smell is supported by other senses (a thesis that will be explored in the next chapters of the book). That's why I urged the Sony team to integrate Aromastic with the mobile phone—it opens the door to more olfactory innovations, games, meditation apps and wine courses. However, if Sony decides to pursue this idea, they will face competition. In recent years, new entrants, such as the Japanese companies Scentee and ChatPerf, have developed ways to spread odors from a mobile phone, using small odor sticks that plug into it. Digital odorizers are already here. But for them to make a real impact, companies need to better understand human psychology—what odor applications will users need and want to pay for?

My research team's contribution to the next generation of high-tech olfactory experiences is NoseWise, an advanced olfactory machine for virtual worlds that was constructed by research engineer Peter Lundén. NoseWise looks like a high-tech spray bottle and is connected to a computer with a Wi-Fi transmitter. The user puts on a pair of VR goggles and is then transported to a virtual wine cellar, where eight wineglasses are lined up. The player—because this is a computer game invented by my research team to train the analytical skills of the sense of smell—chooses how complex the wine smells should be. Then you just pick up a wineglass and smell it. The game itself consists of identifying the different smells in the virtual wine. As in other games, the abilities of the nose can be precisely measured and trained; with different levels of difficulty and keeping track of the score, smelling becomes an exciting challenge.

There is reason to believe that progress in digital smell technology will accelerate in the coming years. The European Research Council recently funded an ambitious research project by olfactory scientists Noam Sobel, Johan Lundström and their colleagues to develop what they call TeleSmell. The researchers aim to develop a system for sampling odors in one location, converting them to a digital code for transmission over the internet and faithfully reproducing them at another location. The project is immensely challenging because of the complex and partly unknown relationship between the chemical odor features and our perceived experience. NoseWise, TTTV, TeleSmell and other smell- and taste-based applications can enrich our digital worlds, so far dominated by image and sound. Perhaps we are at the beginning of a digital smell revolution?

Yes, smells and tastes will continue to play important roles, even in high-tech societies. With the help of new olfactory technology, they will find their way into our smartphones, tablets and computers, and they will probably enrich our lives in ways we cannot imagine today.

•

Smells took a back seat in the modern era. But smells and odors continue to play an important role in our lives. How could it be otherwise? After all, the sense of smell is the most primal of all senses. The history of smell is long, going back almost as far as life on Earth. Over millions of years, evolution has created a variety of advanced senses of smell; they are found in bacteria, fungi, mosquitoes, mice, dogs and humans. Next, we will explore how the sense of smell unites the different forms of life on Earth in our most important common endeavor: finding the next meal.

CHAPTER 3

THE WORLD'S FIRST SMELLS

I DROPPED TO MY knees in the park. I put my nose next to a lamppost and started sniffing. Sure, it looked ridiculous, but I tried to ignore the stares of the other dog owners. Yes, I could smell the odors. Layers upon layers of urine, dirt, metal and grass. I tried to understand what the different smells meant, but soon had to give up the futile attempt. After a moment of crawling, I got up, brushed myself off and moved on. It was an attempt to imagine for a moment how a dog perceives the world. I am not blessed with a particularly sensitive nose, but Nelson, the family's miniature poodle, sniffs his way through life, among lampposts, snowdrifts and trees. He walks up and sniffs, explores, takes a small step and continues. He is blissfully unaware of the alleged division between the thinking and sensory parts of the brain. When he sniffs, he thinks very hard, that much is clear. After a while of sniffing for hidden dog treats, he is exhausted and has to lie down and rest. All dog owners must have asked themselves what is going on in their dogs' brains during these moments. What nuances can they smell and what meanings do the smells carry for them?

It is no coincidence that the dog was the only animal to defeat the human nose in Matthias Laska's competition. The dog can use its nose in a completely different way than is possible for the human nose. Three unique features make the dog's smelling ability superior. (1) The nose is moist, due to the presence of sweat glands and constant licking of the nose. This allows the dog to feel where the wind is blowing—the wind cools the nose as it points exactly in the direction of the wind—and determine where a smell is coming from. Here, the dog is helped by special temperature-sensitive cells on the nose, which make it particularly sensitive to the cooling effect of the wind. (2) The dog's sniffing is different from ours, as the exhaled air is sent out through slits on the underside of the nose. This aids sniffing, which can be concentrated entirely on the incoming air, without mixing in exhaled air. The nose slits send the air to the side, rather than forward. That's why dogs can sniff intensely, without "blowing away" the smell when they exhale. (3) The insides of a dog's nostrils are folded to accommodate as many odor receptors as possible. It is estimated that dogs have 50 times more receptors than humans, and bloodhounds, bred specifically for their ability to smell, may have up to 300 times more receptors. The difference between having 5 million and 500 million receptor cells is obvious to any dog owner who has seen their dog get a whiff.

The sense of smell is something that we have in common with the other animals, because every family in the animal kingdom has a sense of smell. It is the most primal of our senses. And while we may never be able to understand what goes on inside a dog's head, we may be able to understand our kinship with other humans, and with other animals, through the sense of smell. Its history starts early. Really early—at the time of the beginning of life. In fact, primitive smelling and tasting is the basis for all life on Earth. Now take a deep breath through your nose, because here is a brief summary of how smelling and tasting came about.

The very first unicellular organisms appeared on Earth about 3.5 billion years ago, about a billion years after the creation of the planet. This phase is a major part of Earth's history, lasting almost 3 billion years. It was not exactly an eventful time because single-celled organisms were the most advanced thing the planet had to offer. They lived on the carbon dioxide that then dominated the atmosphere— the same molecules that are now found in the air you breathe. Oxygen, which is nowadays vital for us living beings, was a residual product, the garbage left over after the organisms ate their carbon dioxide. Oxygen thus accumulated in the atmosphere, and eventually, a completely new organism emerged, carrying out what could be called the first revolution in world history: it managed to reverse the process and was able to live by converting the abundant oxygen into carbon dioxide. It became the first aerobic—oxygen-using— form of life. Oxygen had an important advantage over carbon dioxide. It provided more energy, allowing aerobic organisms to grow and multiply. This development later led to plants, birds, insects and the earliest mammals. But already during the time of the simplest organisms, the foundation of our sensory experiences was established. And the development was driven by smells.

The aerobic cells made an amazing discovery: they could smell each other! Well, they weren't yet aware of any olfactory sensations, of course, but they were at least beginning to respond to the chemical secretions of other cells—that is, the very earliest form of smelling and tasting. Some of these chemicals were by-products, a sort of prehistoric version of bodily waste. Others contained information that led to a particular reaction—the first chemical signal. Cells had started to communicate with each other.

What does it help a single-celled organism if it has a sense of smell? one might ask. Where it lies in its wet environment, does it matter if it can sense a food source farther away? Scientists discovered the first clues to answer these questions back in 1881, when German biologist Theodor Wilhelm Engelmann found that bacteria seemed to be able to move in the direction of an energy source. Engelmann's observation created a whole new field of research: chemotaxis—chemically controlled movement. Chemotaxis is something that many simple organisms can do. Bacteria like *E.coli* have flagella, a kind of tentacle that they use to move around. Using the flagella, the bacterium can move in two different ways, either spinning around in the same place or swimming forward. Unfortunately, *E.coli* has no eyes or brain, so it doesn't really know where it's going. But when it smells an attractive odor, it starts swimming straight ahead—no matter where the odor comes from. If *E.coli* swims in the wrong direction, the attractive smell becomes weaker. Then the swimming stops, and the bacterium starts spinning instead, and then takes off in a new direction, as randomly as the first. If it manages to swim in the direction of the source this time, the swimming continues because the chemical signal does not vanish, and the food source is finally within reach. With a few simple movements, and without any thought, *E.coli* can navigate its environment using its sense of smell. Scientists have now discovered a variety of such movement patterns in the simplest of organisms. Recently, Belgian scientists managed to re-create such a mechanism by putting a bunch of bacteria on a starvation diet—something they thought would trigger new evolutionary mutations. Sure enough, after a while the bacteria mutated in such a way that the sense of smell started to control the flagella, and the bacteria became mobile, just like *E.coli*. So the next time you walk past a hamburger restaurant and get caught up in the smell of fried meat and fries, consider that you are re-creating one of the world's oldest, and most important, behaviors: the ability to smell your way to the next meal.

Primitive smelling and tasting gave rise to more advanced animals. The sense of smell is really about individual cells in the body reacting to chemical secretions. And a cell's environment inside the body is . . . other cells! The body's different cells learned to communicate with each other inside the body, using chemical secretions.

The "internal sense of smell" was the starting point for the biological evolution that led to humans and other living animals. As the philosopher Peter Godfrey-Smith described it, animal evolution began when cells "gave up their individualism" and became part of a larger machinery of increasingly interconnected cells. The single-celled organisms at the beginning of life had interacted with their environment using chemicals—what modern people would call smells and tastes. The same chemical communication that took place between single-celled organisms became a natural starting point for the internal signals of multicellular organisms. Thus, about 0.6 billion years ago, a second revolution in life on Earth took place. A cell division, the process by which a cell is duplicated by copying its chromosomes and then splitting into two new cells, was half successful. The two resulting copies of the cell failed to separate. But they still somehow managed to survive within the same outer shell, and the result was a larger, multicellular organism. Scientists believe that the first multicellular creatures were small aquatic animals, the precursors of today's marine sponges and jellyfish. The larvae of sea sponges travel in the water and smell their way around. Our bodies, like those of sea sponges, function thanks to chemical communication between cells. It is like a symphony of millions of smells and tastes inside our bodies. Just as it has been since the early days of life on Earth.

Not only is the chemical sensing of simple organisms of historical interest, but our knowledge of their chemical communication can actually help us create a better life here on Earth, right now. Take fungi, those amazing organisms that are neither plants nor animals. Fungi have a great ability to smell their food. But unfortunately, their food is often rice or other human food. Fungus attacks destroy food for 80 million people every year. Turning off the fungi's sense of smell with genetic engineering may be a future method of preventing crop failure and famine. In other cases, we may want to improve the fungi's sense of smell. Researchers at University of California, Riverside have discovered that one type of fungus, *Neurospora crassa*, smells the cellulose it eats. Its spores pick up chemical substances outside the cell and this sets the process in motion. The fungus eats the cellulose and converts it into glucose. *Neurospora crassa* could be of great benefit to humanity. They do not compete with us for food, as cellulose is not something humans eat. Researchers are now trying to learn how to control the fungi's sense of smell to process the cellulose. If they succeed, fungi could become our best allies in producing new environmentally friendly fuels for cars and buses.

The animal whose sense of smell is most important for us to understand is probably the mosquito. Some of us are "mosquito magnets." This is because our body odor attracts mosquitoes. This is a major source of irritation and many people speculate on why they have such "sweet blood." Some say that it is the blood type that determines the interest of mosquitoes, or that they can be fooled by our eating garlic or vitamin B. However, there is no scientific support for these theories. We still know very little about how mosquitoes select their victims. But studies are ongoing, because this is one of the most important questions in olfactory research. Mosquitoes' sense of smell has enormous consequences for humans. The mosquito spreads viruses among humans when it sucks blood, particularly affecting Central African countries and other countries near the equator. The odor-sensitive *Aedes aegypti* mosquito spreads the Zika virus and dengue fever, and the *Anopheles* mosquito spreads malaria, one of the world's deadliest diseases. New research by Leslie Vosshall and her colleagues at the Rockefeller Institute in New York shows that the "mosquito magnets" have particularly high levels of the odorant carboxylic acid in their sweat, compared to those that the mosquitoes reject. The researchers collect the smell of sweat through nylon socks worn by the participants on their arms. They then let the mosquitoes choose between the socks to see which sweat smell is most appetizing to them. Using careful genetic engineering, the researchers can then knock out the mosquitoes' odor receptors and test whether this affects their choice. In this particular study, they found an olfactory receptor that appeared to be necessary to guide the mosquitoes in their choice—a receptor that was sensitive to carboxylic acid. Could this research lead to the construction of "odor traps" for malaria mosquitoes and other super-spreaders? So far, these odor traps are not very effective, because even the mosquito's sense of smell is very complex, but the more we learn about it, the closer we get to solving the serious viral diseases that plague warmer countries.

Human chemical senses, like those of fungi and mosquitoes, work through microscopic receptors that respond to specific molecules in the environment. In humans, these receptors are located on special smell and taste cells in the nose and mouth. Their job is to translate the chemical environment into activity in the nervous system. This activity consists of small, rapid electrical signals that are sent to the brain, where they are interpreted. But odor receptors are a fairly recent discovery. Until 1991, the biology of the sense of smell was considered something of a mystery. No one understood how the sense of smell translates chemical signals in the environment into nerve impulses that could result in so many different scent experiences. The big breakthrough came when Linda Buck and Richard Axel at Columbia University in New York published their groundbreaking results on the receptors of the sense of smell. Buck and Axel researched mice, the animal that has become something of a model for how the mammalian brain works. When Buck and Axel mapped the "family" of genes that create the olfactory receptors, the results were astonishing. At the time, other such families were known to be responsible for producing different receptors in the body. The olfactory receptors are of a type that usually sit on the outer shell of cells and communicate with the environment. Such receptors are of great interest for research into new drugs. After all, the goal of drug research is to create new molecules that can affect the body's cells—and to do that, you need to reach the receptors that manage the cell's communication with the outside world. Olfactory receptors are long proteins that are folded into the outer shell of the nerve cell. The receptors wind their way in and out of the shell. This forms a pattern that can be described as a kind of keyhole. Only odor molecules that fit well into this keyhole can activate the cell. The mice's olfactory receptors shocked the researchers: they had one thousand different receptors—a full 3 percent of the mouse's gene set was devoted to creating them alone. The largest previously known family of receptors had only a dozen members. This was clear evidence that the sense of smell is one of evolution's most important inventions.

The mouse has a sense of smell that is similar to humans. According to Laska's comparison of odor sensitivity, humans and mice came out pretty close: 36 to 35. But that's not the whole story. The mouse actually has four different senses of smell. In addition to the normal sense of smell, which is comparable to ours, there are three other olfactory organs, in and on the nose, which seem to have special functions. These allow the mouse to detect pheromones secreted by female mice in their urine, and warning signals such as the smell of foxes and cats. Mice live in a world that is completely dominated by smell and most of the important events and other animals in their lives are reflected in their four senses of smell. Simply measuring olfactory sensitivity is an oversimplification—the mouse's sense of smell is probably richer than ours, in ways we don't yet understand.

Buck and Axel's discovery demonstrated that the ability to perceive odors had been prioritized by evolution, and that olfactory sensations arose from the creation of a very large group of specialized cells to respond to chemical substances in the environment. The sense of smell was now back on the big stage of science. After Buck and Axel's breakthrough in 1991, many thought the mystery of smell was solved. When we have so many different types of receptors—humans have about four hundred active receptor types—you might think that every odor molecule we can smell has its own specialized receptor, and that there is a simple link between molecule and receptor. But this was a premature conclusion. It soon turned out that the way in which odor molecules attached to the receptors was not quite as simple as had been thought. Each receptor can capture odor molecules based on several different chemical properties. And to make things even more complicated, each molecule can attach to several different types of receptors. Add to this the fact that many of the smells we experience in our daily lives, such as coffee, flower scents or exhaust fumes, are made up of dozens of different molecules, and we can start to realize just how sophisticated our chemosensory world really is. So far, we only understand a fraction of it.

The original role of the sense of smell and taste, going back billions of years, is therefore to find food. Throughout most of evolution, food has been scarce, and only those who have prioritized their own energy supply have been able to survive and reproduce. No wonder we humans are so good at eating and drinking. But our brains are not adapted to have access to high-calorie food at all hours of the day. We love sugar, and even newborn babies smack their lips happily when given a small sip of sugar solution. Evolution has given us an innate attraction to energy-rich foods. But these innate preferences sometimes lead to problems. In 2016, a line was crossed. That year, for the first time, more than half of Swedes aged 16 to 84 were reported to be overweight. Obesity has many causes, but the senses of smell and taste, and their close links to the brain's reward system, play an important role as they create impulses to eat. We indulge in food that we know is unhealthy and we often eat too much of the good stuff. Most people have to live with the tug-of-war; we oscillate between being guided by our impulses and our ability to control them. Indeed, our unique senses of smell and taste may partly explain why some of us become overweight.

Some people may have an extreme sensitivity to smell, while others are insensitive to the same odorant. The difference can be huge. Comparing people who appear to have a perfectly normal sense of smell, it can sometimes take up to 100,000 times more odor molecules for the most insensitive person to detect an odor that the most sensitive person in the group can smell very easily. In other words, if it takes ten molecules for you to detect a certain smell, it may take a million molecules for your friend. So it's not surprising that we easily get into arguments like "Well, can't you smell it?" The extreme differences between people are partly evened out if we compare many different odor molecules. Someone who is insensitive to one molecule may be sensitive to another. Fortunately, nature's smells usually come to us in a mixed form. So even if you lack the ability to smell one of the molecules involved in the scent of freshly brewed coffee, you can probably smell the others.

The large variations also apply at the neurobiological level—as much as 30 percent of olfactory receptors differ on average between two people. This means that our brains do not receive the same signals, even if we smell the exact same odor. And this variation can make a big difference. An Icelandic research team recently discovered an unusual gene variant that affects the perception of fish smell. They asked thousands of Icelanders to rate different smells and matched the results with the participants' gene variants. The smells included trimethylamine, which is found in fish, and at particularly high levels in rotten or fermented fish. The people with the most common variant of the gene perceived the fishy smell of trimethylamine as disgusting and described it as "rotten." But Icelanders with the rare gene variant found the smell more acceptable and used more neutral words to describe it. The rare gene variant may have been particularly beneficial in Iceland, where fish have long been the dominant source of protein. A traditional Icelandic dish is Hákarl, fermented shark, which, like Swedish surströmming, has an intense smell, but is nevertheless appreciated by some Icelanders. Perhaps it is the unusual gene variant that makes some Icelanders like the fermented shark. Research shows that the variant is particularly common in Iceland. Tolerating strong-smelling fish may have benefited those who lived during periods of poverty and lacked a varied diet.

The variability of the sense of smell can have consequences for what we eat and drink, and thus the sense of smell can also affect our weight. The smell of high-calorie foods is often perceived differently by people who are overweight. David Garcia-Burgos, a researcher at the University of Granada, says that obese people are particularly good at distinguishing the smells and tastes of foods with different calorie content, such as products with different levels of fat. This means that an obese person may, for example, find that a light yogurt tastes worse than a higher fat yogurt, while a normal-weight person finds the taste equivalent. As you now know, the sense of smell often plays a major role in tasting food. So the sense of smell and taste can make us like or dislike different products. But there is other research that points in a different direction. Sometimes overweight people can actually have a poorer sense of smell than standard-weight people, and this can also have negative consequences. On average, people who are overweight think that wine has less taste than a standard-weight person does. According to some researchers, this is because overweight people's saliva contains specific chemicals that reduce the odorous substances released in the mouth and that travel to the nasal cavity via the throat. This has led to the theory that higher consumption and obesity may at least partly compensate for reduced sensitivity to taste and smell. The less you experience smells and tastes, the more you need to drink to get the same experience. What to think—do overweight people have better or worse senses of smell and taste? The research is in-conclusive and both theories can be true: obesity can have different causes for different people.

The senses of smell and taste are controlled by the body and its urges. Hunger can make us all notice food smells that we would otherwise ignore. The senses of smell and taste are strongly intertwined with our brain's reward system. Throughout our lives, we encounter smells that we associate with strong positive emotions, and these smells seem irresistible. They trigger cravings, which can be measured as a signal in the brain's reward system. When we encounter the smell of a grilled hamburger, a sort of tug-of-war occurs within us. Ohio State University researcher Dylan Wagner specializes in studying how this tug-of-war plays out inside our brains. In one experiment, he had smokers and nonsmokers watch movies with lots of smoking, while lying in a brain-imaging scanner that continuously measured brain activity. He found that the smokers' reward system increased in activity precisely at the times when they saw others smoking. Their brains now began to bring back memories of the taste and smell of cigarette smoke. But the brains of the nonsmokers did not react at all—for these people, no cravings were aroused. The reward system works similarly for different kinds of temptations. This is why some people are often at risk of falling into multiple addictions, whether it's gambling, smoking, alcohol or food. The reward system reacts strongly to these different kinds of temptations, and the brain's so-called control system is unable to tame the impulse. This may further explain why some people become overweight—they have a particularly high signal in the reward system. Ideally, someone with a highly active reward system should also have particularly developed neural pathways between the brain's reward and control systems—this seems to favor the control system's power to curb eating and drinking impulses and is associated with a reduced risk of becoming obese. By measuring such signals in the brain, researchers can even predict how much alcohol young people will drink when they enter college life, or how much weight people will gain in the future.

We are born with an affinity for sweet flavors, but we also have an innate tendency to avoid strong bitter flavors. For example, have you ever tried to get a baby to drink espresso? Don't do it—you're wasting a good cup! In research experiments, babies who are given a taste of bitter water (a caffeine-free, harmless and more child-friendly alternative to espresso) immediately spit it out and show with very clear grimaces that this was really disgusting. This innate reluctance is probably due to the fact that many of nature's toxic substances taste bitter. Thus, it has paid off for the survival of the human race to want to spit out things that taste bitter. But different people have different sensitivities to bitter tastes, and how this works has long been debated among scientists. The debate started back in the early 1900s, when chemists learned how to make the substance creatine; the chemists tasted this new substance (a common procedure at the time) and much confusion ensued. Some thought it was tasteless, while others thought it had a strong bitter taste. How could they react so differently? The confusion among chemists led scientist Arthur Fox to investigate the substance PTC in 1931, a substance that also proved to divide the population into two camps. They had no idea why only some people could feel the bitter taste, so they tried various experiments, one of which was to divide people according to their sensitivity to PTC. These groups of "tasters" and "nontasters" were then asked to rinse their mouths with each other's saliva. The researchers' hypothesis was that it was differences in the taste of the saliva that influenced an individual's susceptibility to PTC. The experiment was not only quite disgusting, but also proved to be a scientific dead end. It is not our saliva that makes us sensitive to the bitter taste, but the receptors on the tongue.

Soon after Fox's discovery, other researchers began comparing children and parents in different families and found that bitterness sensitivity was inherited—children's sensitivity to bitter tastes was similar to their parents'! This led the researchers on to the right track: our genes help produce different receptors in the body, which in turn mediate different taste sensations. The discovery had unforeseen applications, as for a time it was used to establish paternity. When there was disagreement about who the father was, a simple test was devised. If both the mother and her partner were "nontasters" (unable to taste PTC), then the child should be too, as the child would have all the genetic conditions for the same taste sensitivity. But sometimes the child would spit out the drink—that must mean that someone else was the father!

Scientists are still debating how much our individual genetic makeup influences what we like about bitter food. Indeed, bitter taste is not universally disliked. Many things we humans consume taste more or less bitter, such as broccoli and coffee, but also cigarette smoke and alcohol. One theory is that those who have specific gene variants that make them sensitive to bitter substances also tend to avoid vegetables and other bitter flavors. They simply don't like the taste because they perceive it as too bitter. Researchers have argued that the PTC sensitivity gene determines how many bitter-sensitive receptors you have on your tongue, and the more you have, the less you like bitter flavors. Does this mean that everyone who likes milk or sugar in their coffee is a "taster" and those who like alcohol are "nontasters"? No, unfortunately it is much more complicated than that. In fact, there are about thirty genes that affect sensitivity to different bitter flavors. Most people are sensitive to some of the bitter substances, but insensitive to others. Coffee has two different types of bitter flavors, so it is conceivable that a person is sensitive to one but insensitive to the other. It has therefore been difficult to link individual gene variants to specific eating habits.

An exception to the rule is how we perceive the smell of cilantro. Are you one of those who opt out of fresh cilantro when eating Mexican food? You're not alone—about a fifth of the population finds its taste and smell soapy and unappetizing. About ten years ago, researchers discovered that there was a genetic reason why people have such mixed feelings about cilantro. They found two genes that differed between people who thought cilantro tasted like soap and those who liked it. The two genes were found to be responsible for creating odor receptors for a specific type of odor: aldehydes. So these two genes largely determine whether you love or hate cilantro.

If the genetics of the sense of smell differ so much among individuals, does it also differ among different populations? This question has not yet been clearly answered, but there is some evidence of such group differences. Icelanders' tolerance for fishy smells is a recent discovery. Tolerance of aldehydes, a chemical family of fruity and pleasant scents, can also reveal regional differences. Aldehydes are widely used in perfumes and in detergents. But some aldehydes are intrusive and pungent (formaldehyde, the liquid used to preserve brains, is one). And different people react very differently to aldehydes. In a study based on New York City residents, participants of Asian origin found the smell of aldehydes stronger, compared to Europeans, who found the aldehydes less strong. The difference may be genetic. In fact, in some Asian countries, aldehydes have been reduced in several household products because it has been found that people find aldehyde odors strong and unpleasant. Are the differences due to the fact that Asians and Europeans have partially different smell receptors? It is possible—but no such differences have yet been proven.

Heredity and environment—both shape our eating habits, and despite thousands of published research reports, it is often difficult to separate their influence. Although differences among people *within a group* are often largely due to genetics, differences *among groups* tend to be mostly due to cultural reasons. A good example of this is obesity, which is more common in the US than in France. At the same time, there are obese and thin individuals in both countries. Is obesity hereditary or environmental? Research suggests that differences between fat and thin individuals *within* a country are mainly due to the genetic lottery. Some Americans, and some French people, are born with genes that allow them to gain weight more easily than their compatriots. But the difference *between countries* is not due to genetics but to cultural factors. Americans are served larger portions and more calories. They eat more often alone, in front of the TV and in the car, which means they don't think as much about the smell and taste of food. The brain does not perceive that the reward system has stopped signaling—we are full—and eating continues. These are all cultural factors that increase the risk of obesity.

Genetics and culture have long been known to influence the way we eat, but these explanations still have limited explanatory power. A third explanation is now receiving increasing attention. This is the cognitive perspective, which has become increasingly important in our understanding of eating and obesity. Researchers have found that it is not only our external sensory impressions and internal feelings of hunger that attract us to eating. The ability to imagine food is an important cause of strong food cravings. Our mind alone can create a rich internal image of a juicy hamburger. We can imagine the meat, the cheese and the other toppings, we can feel the soft slices of bread in our hand, and how the rich smells and tastes wash over us. Such a little thought exercise can actually be enough to make some people start looking for the nearest burger restaurant. One sense that stands out here is the sense of smell. Almost everyone can imagine what the hamburger looks like, but the ability to bring its smell to life is unevenly distributed in the population. A good ability to imagine smells can have particularly important consequences for our eating habits. A research team led by Dana Small at Yale University has shown that the participants who are best at imagining smells are also the most likely to be obese. The researchers' theory is that it is the vivid mental images of smell that create strong cravings for fragrant, high-calorie food. In a recent study, they also measured the brain activity of participants who were asked to imagine the smell of cookies, roses and other smells we encounter in everyday life. Using an advanced statistical analysis, they were then able to classify the brain patterns and identify which smells the participants were thinking about. The participants who had the strongest ability to imagine smells also had brain activity that most clearly indicated which smells they were thinking about. A kind of olfactory mind reading with modern brain-imaging techniques. The cognitive perspective on the sense of smell may thus provide the missing piece of the puzzle that allows us to gain a complete understanding of why so many of us have strong cravings for high-calorie food.

The sense of smell is a primordial sense. It is present everywhere in the animal kingdom and provides an insight into how the animal brain works. Only in recent years have scientists understood the complexity of the sense of smell. At the same time, it is not a mass-produced product, but rather a "limited edition," with special characteristics just for you. It reflects not only your genetics but also your history, culture and desires. The cognitive perspective can help us understand how desires arise in our brains and make our favorite foods so irresistible.

PART II

THE SMART NOSE

CHAPTER 4

THE EMOTIONAL TIME MACHINE

WHAT IS YOUR favorite smell? This question may seem banal, but think carefully, because your answer will say something about who you are. Our most meaningful smells evoke intimate feelings, similar to how we are affected by music or art. It is almost impossible to talk about your favorite smell without getting personal. The smell of our partner or our children evokes feelings that are difficult to describe in words. A friend who recently became a father described the smell of his child as making him giddy with love. For dog lovers, the smell of the family dog is often high on the list of favorite smells. These feelings are shaped by our memories and experiences. Researchers have shown that people often think other people's pets smell bad, but not their own pets. New parents have the same attitude toward dirty diapers—it doesn't smell as bad if it comes from your own child. These smells have a special meaning for us, and the emotions they evoke are due to olfactory memory—smelling enables us to remember events and emotions, and associate them with each other. This is what makes smells so deeply personal.

Like few other impressions, smells can take us back to childhood and bring back old memories. It is an "emotional time machine," according to olfactory memory researcher and my long-term collaborator Maria Larsson. Most of us have one or more smells that evoke childhood memories. The smell of lily of the valley, for example, can remind us of the bouquet of flowers we picked that special last summer before starting school. This peculiar phenomenon is usually associated with the writer Marcel Proust, who, in the first volume of *Remembrance of Things Past*, dips a madeleine cake in lime-blossom tea. It is a very special novel, not only because of its length, seven huge volumes, but because of the winding, associative style with which Proust demands the reader's attention. The focus is on sensory impressions; it is the senses that evoke feelings, thoughts and reflections. And which bring back memories of childhood.

Proust begins by describing how for a long time the protagonist could not remember more than the most basic and general aspects of his childhood in Combray—the name of the village in which the book's protagonist grew up. The familiar surroundings were hidden in "unchanging evening light." Proust writes of the past as an elusive world that we cannot summon with our minds. But everything changed with the madeleine cake his mother served him with tea on a cold winter day. . . . *No sooner had the warm liquid mixed with the crumbs touched my palate than a shiver ran through me and I stopped, intent upon the extraordinary thing that was happening to me.* This is followed by an intense passage about how the memories begin to haunt him, announcing their presence for brief moments and then sinking back into the darkness of oblivion. He tries intensely, seemingly at the cost of his life, to get hold of the elusive memories. Finally, with the help of smells and tastes, he manages to unlock the treasure chest of memory. In Proust's hands, sensations and associations become great literature.

The sight of the little madeleine cake had recalled nothing to my mind before I had tasted it; perhaps because I had so often seen such things in the meantime, without tasting them, on the trays of pastry cooks' windows, that their image had dissociated itself from those Combray days to take its place among others more recent. . . . But when from a long-distant past nothing subsists, after the people are dead, after the things are broken and scattered, taste and smell alone; more fragile but more enduring, more immaterial, more persistent, more faithful, remain poised a long time, like souls, remembering, waiting, hoping, amid the ruins of all the rest; and bear un-flinchingly in the tiny and almost impalpable drop of their essence, the vast structure of recollection. . . . And as in the game wherein the Japanese amuse themselves by filling a por-celain bowl with water and steeping in it little pieces of paper which until then are without character or form, but, the mo-ment they become wet, stretch and twist and take on colour and distinctive shape, become flowers or houses or people, solid and recognizable, so in that moment all the flowers in our garden and in M. Swann's park, and the water lilies on the Vivonne and the good folk of the village and their little dwell-ings and the parish church and the whole of Combray and its surroundings, taking shape and solidity, sprang into being, town and gardens alike, from my cup of tea.

For quite some time, Proust was alone in exploring the connec-tions between olfaction and memory. When researchers ask their study participants to write down their notable memories, they tend to focus on experiences that they can visualize and that have had a significant impact on their lives: a childhood friend, an early romance or perhaps negative experiences of victimization and bullying. The more vivid the memories, the more activity they elicit in the brain's visual cortex. When studying memory in this way, smells are rarely included. However, another memory method began to be used in the 1970s, which paved the way for the study of olfactory memories.

This method used a list of words presented to all participants as memory cues: *library*, *clock*, *factory*, *street* and so on. The words, researchers would later realize, could easily be exchanged for different sensory impressions, enabling a comparison of how these impressions triggered different types of memories. Among the first to use this method to study olfactory memories were Maria Larsson and Johan Willander, who invited hundreds of elderly people to their research laboratory at Stockholm University. They asked participants to describe the memories triggered by words, sounds, images—and smells. The results were surprising. Smell memories were dramatically different from memories triggered by other cues. It was known that words trigger memories from two specific periods of life. One period is the last few years. A third of all memories evoked come from experiences in recent years, which is not surprising as they are particularly fresh in our minds. However, we tend to forget events that are farther back in time, when more has been lost to oblivion. The most interesting period is therefore not the one that is closest in time, but the second period, where many memories are evoked despite being far in the past: adolescence and early adulthood. Here we see a curious "bump" in the memory curve that extends roughly from age eleven to thirty and peaks at about age twenty. This period of time sees an unusually high accumulation of memories, despite being sandwiched between two other periods that we don't remember very well at all, childhood and, if we are old enough to have passed it, middle age. The experiences of childhood are believed to be inaccessible to us because our brains are simply too immature at that time to create lasting memories of our experiences. Our knowledge of the world, our identity, self-image and language skills help us remember our lives. But in childhood, these skills are far from fully developed, and as a result, childhood experiences are often not recorded in memory. So researchers have developed theories about the "bump," the accumulation of memories of our experiences during adolescence and early adulthood. Some argue that it occurs at this time because the brain's memory capabilities are at their best. Others argue that it is a particularly eventful time of life that includes the entry into adulthood, where crucial life choices are made and where one's future life begins to take shape. Perhaps both theories are true. But the strange thing about olfactory memories is that they do not come from the same time period as the other memories. The olfactory bump appears already at the age of six to ten years.

It is now well established that smells have a unique capacity to evoke memories of childhood. But that is not all. Olfactory memories are also different in other ways. They evoke a sense of "going back in time" in a way that no other sensory experience does. Smells can make you relive the time you helped your grandfather tar the rowing boat, the time you stepped out of the airplane on your first vacation to the Mediterranean or the wedding where you got to be a bridesmaid with your own bouquet of flowers. You remember the details, relive the moods. Smell memories are also more emotional than other memories. But why do smells have this property? It is still something of a mystery, hidden in the twists and turns of our brain. And what makes it even stranger is that our memory for smells is not really very impressive in itself. When my colleagues and I examine the memories of different experiences under controlled conditions, we consistently find that our participants are much better at remembering images, words or sounds than smells. But even so, some of the smells that the brain stores are retained, and sometimes we carry them with us throughout our lives. They are more faithful and enduring, as Proust would have said. One reason for this could be that the different sensory impressions activate different brain areas, thereby engaging different psychological memory processes. In one experiment, participants smelled smells and listened to words that evoked childhood memories, while their brain activity was recorded. The results showed that both the smells and the words triggered activity in similar areas associated with memories, emotions and visual impressions—a sign that the cues triggered stored memories that participants could "see" in their minds. But it was only the words that also evoked strong activity in the frontal lobe, where language and abstract thinking reside. The smell memories did not. One hypothesis is, therefore, that words can trigger more abstract thinking. With words as clues to memory, we can search our memory bank in a more systematic but perhaps also more impersonal way. Words can help us find more memories. But for smells, this strategic option does not exist—smells can only bring to life the personal experiences, those that have a clear sense of personal presence and emotional charge. My own research supports this interpretation. My colleagues and I have studied the olfactory brain's connection to the brain's memory center, and compared the strength of this connection to hearing, sight and touch. This is measured by looking at how the activity in the areas increases and decreases; if two areas "oscillate" in sync with each other, it can be interpreted as close communication. And it turns out that the olfactory brain, more than the other senses, oscillates in time with the brain's memory center. But the sense of smell's direct contact with the memory center probably comes at a cost. Smell does not have the same ability as do words to inspire abstract thought and the creation of strategies for memory retrieval. This leads to fewer but more concrete, intense and emotional memories being evoked by smells.

But a mystery remains. Why do the smells evoke such old memories? So far there is no definitive answer to that question. But my research with colleagues at KTH Royal Institute of Technology in Stockholm provides a possible explanation. We already know that the sense of smell develops very early in life. By mid-pregnancy, the fetus can smell the food the mother eats. French researchers have been able to demonstrate this using aniseed pastilles, which they allowed pregnant women to eat in large quantities. When the babies are born, they seem to be particularly interested in the smell of anise—they turn their heads toward the smell of anise in the cradle! The smells a baby is exposed to in the womb can actually influence what food they will be interested in in the future, so strong and so early are our first olfactory memories. And we believe that this early development of the sense of smell can explain the memories that Proust described so vividly. Our calculations show that if the olfactory brain "matures" earlier in life than other brain systems, this can actually explain the childhood memories. Nerve cells become less and less receptive to new memories as they get older. This means that the malleability of the sense of smell may be greatest in childhood. Later, when the malleability has decreased, the childhood smells that we have memorized are protected because the neurons are no longer as malleable—but this comes at the cost of the smells we encounter in adulthood not being as easily stored. And this is exactly what we find in our memory experiments: adults often have difficulty remembering new smells. Other sensory systems mature later than the sense of smell, which may explain why, for example, visual and auditory memory storage is strongest in adolescence and early adulthood, but smell memory is not. So perhaps we are on the way to solving the mystery.

Memories are the most important building blocks of our lived experiences. What we remember becomes part of us, of our identity and self-image. Memories allow us to develop, both as individuals and as part of society. But memories are not just a mechanical recall of past experiences. When we remember, we also create a story about ourselves. Your olfactory memories are unique, but they probably reflect something that is still important to you today. I am often struck by the way smells are used in autobiographies. Despite the fact that smells are usually so difficult to describe, skilled writers can use them to engage the reader emotionally. Describing smells, like few other impressions, can bring a presence to the text, giving it a personal appeal. Märta Tikkanen's *Love Story of the Century* is about a destructive marriage with a drunken, abusive husband. She describes how the smells of cigarettes, sherry and stale beer take over the family's apartment and permeate their sad existence. Other descriptions of smells are less graphic, but still reveal strong emotions. Eric Rosén writes about his alcohol-dependent father in the book *Jag ångrar av hela mitt hjärta det där jag kanske gjort*. Rosén describes without frills the smell of spoiled food from the refrigerator that has been turned off because his father has not paid the electricity bill. No graphic descriptions are needed to make the reader feel sad and angry that children have to grow up like that—the smell explains everything.

Literary descriptions of smell can also tell more uplifting stories. The Swedish physician and professor Hans Rosling's autobiography *How I Learned to Understand the World* opens with not one but two powerful olfactory memories. One is the memory of coffee-smelling coins that Rosling's father brought home from work at the Lindvalls coffee roastery in Uppsala. The workers who packed the coffee in South America put the coins in as a greeting to their colleagues on the other side of the world, and Rosling's father told his son about the distant countries where the coins came from. This sparked Hans Rosling's interest in the world and global developments. The second smell that opens Rosling's autobiography is less pleasant—the smell of sewage. He remembers the smell from his childhood, when the sewers were still open and ready to spread disease. Sewers were covered in Sweden in the 1940s (an important part of modernization), but Rosling describes how he often visits countries that do not yet have a functioning sewer system. It brings back memories of his childhood, but for the eternal optimist Rosling, the smell of sewage also evokes a sense of hope, because more and more countries are now reaching a level of development where the spread of infection and child mortality can be minimized.

Exploiting Proustian olfactory memories—especially nostalgic ones—is every marketing expert's dream. Anyone who can unlock the door to our childhood with a simple smell has a world to gain.

But olfactory memories are unique to us. No two memory portals are the same. Each person must therefore have their own olfactory key. And what unlocks each portal can smell like pretty much anything. On April 1, 2019, the Swedish Scouts launched their own fragrance, Go Camping, inspired by camps, summer, outdoors and nature: "A deep base note of campfire is tastefully combined with middle notes such as pine and leather, and top notes of dewy summer meadow . . . and damp wool socks." An April Fool's joke, of course, but surely anyone who has been a scout, or had similar nature experiences, can also enjoy the smells that make others wrinkle their noses.

Now I want to ask you to do a little experiment. Ask someone to take three items out of your pantry or refrigerator and hold them under your nose. Close your eyes, and try to name the smells they contain without any other clues. How did it go? You probably got one of the three right. This is exactly what olfactory researchers discover when they ask their participants to name everyday smells. Most people are amazed at how difficult it is to recognize familiar spices such as thyme, allspice or basil. They often say that the smell is very familiar, and they can often provide some information about the smell. That it is a spice, a fruit, a flower. But the exact source of the smell often remains inaccessible to us. If we have such an excellent sense of smell, shouldn't we be better at recognizing and describing smells?

Smells seem to have no language of their own. Compare with colors—we usually know exactly how to describe colors, and we usually agree with each other on what color an object is. Color names are abstract terms that can be used to describe a variety of objects. The word *red* is used to describe the color of a rose, car, tomato—or Uncle Ivan, who forgot his sunscreen. With smells, we struggle to find the words, and when we do find the words, they are concrete comparisons. We say, "This smells like a rose." The same pattern exists all over the world; smells are described by their possible "sources"—

what is it that smells like this? The only abstract categories that all the world's peoples seem to agree on are that some smells are pleasant and others are disgusting—and what is pleasant and what is disgusting may differ between persons and cultures. The difficulty of naming smells is probably one reason why the sense of smell has so often been undervalued throughout history. If people cannot talk about their olfactory experiences with others, they risk being forgotten. The difficulty of describing smells in words contributed to the fact that doctors stopped practicing smell-based diagnosis of diseases. However, I am still often told by experienced nurses and doctors that they can smell what disease their patients have. (Disease smells will be described in more detail in later chapters.)

Can brain research explain why we are so bad at naming smells? After all, hearing and vision seem to be well suited to linguistic communication. We can read, listen and speak. But if our sense of smell is so good, shouldn't we be able to talk about smells as well as we can talk about art or music? Different researchers have different theories. My own research has led me to believe that the answer lies in the link between the sense of smell and the language areas of the brain. The olfactory brain has an older, simpler structure than the visual brain, which sorts visual impressions based on where in the visual field they occur, and processes them in specialized areas for shapes, words, movement, places, colors, etc. The olfactory brain has no such support. As a result, we tend to experience smells as a whole, rather than by identifying individual details. A smell, with its many different molecules, cannot be disassembled into its components, while music and images can be analyzed in tones and colors. This lack of detail makes smells somewhat vague and indeterminate, and makes it difficult for the language system in the brain to put into words exactly what the smell is.

It may seem paradoxical that we can't name smells even though we are so good at detecting them, but this could actually be a key to understanding how the human sense of smell is designed, and it is one of the cornerstones of the cognitive perspective on the sense of smell. I studied this phenomenon alongside the innovative olfactory neuroscientist Jay Gottfried at Northwestern University. Together, we discovered that people are surprisingly quick to match smells to written labels such as *lemon* and *almond*—it takes less than a second. I was impressed by how the olfactory brain makes such effective use of information from other senses.

When we use our sense of smell in everyday life, we rarely have to close our eyes and guess. Instead, we can rely on cues from other senses. Taking cues from other senses is one of the most important features of the sense of smell. It often happens automatically. If we open the fridge after being away and are struck by a pungent smell, we can easily guess that the milk has gone sour. But if we were to encounter the same smell while walking in nature, we would probably be confused instead. What is it that smells so sour . . . juniper bushes? Every environment we find ourselves in comes with associations and expectations about what to anticipate next. For our brains, this is a smart solution: by creating expectations based on our sensory impressions, the brain helps us to behave appropriately in the next moment. This is something that all the senses use. When you're out for a walk in the woods, you automatically lift your feet to avoid tripping. When you see a color, your brain adapts the hue to the background light— the color you experience is a kind of brain guess, and many optical illusions are examples of how the brain can guess wrong. The brain's guesses and adaptations are largely automatic, outside our consciousness. The sense of smell works the same way, but to an even greater extent than the other senses. The environment you are in will, consciously or unconsciously, set you up to smell different smells. If you enter a new environment, such as a friend's apartment, you will sniff as you cross the threshold. What does the new place smell like? The sense of smell rushes to form an overall impression with the help of its fellow senses: our eyes, ears, hands or taste buds. But what makes the sense of smell unique is that it needs the help of other senses for identification. Here we approach one of the most distinctive features of the sense of smell. The sense of hearing or sight usually manages on its own. However, the strong suggestibility of contextual cues is one of the distinctive features of the sense of smell; it is the key to its richness, but also its limitations. The realization that the sense of smell is built for interactions with the brain's thought and emotional systems, and with the other sensory systems, is crucial. That is the reason why smells may influence our eating habits, our attitudes and prejudices, and the interactive properties of smell may provide a key to helping those who have lost their sense of smell to regain it. The special nature of the sense of smell means that those of us researching its psychology and brain mechanisms can constantly find new and exciting results—if we dare to broaden our perspectives. And we will do so in the next chapters of this book.

There are examples of groups that have a well-developed language of smell. Cognitive scientist Asifa Majid of Oxford University has studied the Malay people, hunter-gatherers who live in the rainforest areas of Thailand. In the dense forest, it is often dark and you cannot see very far ahead. For those who make their living there, it can be useful to have a keen sense of smell, which can detect the scent of edible plants or threatening predators from afar. The hunter-gatherers studied by Majid also have a different way of communicating about smells than we do in the West. Their language has specific words that only describe smells, such as the word *cŋes,* which is used to broadly describe the smells of gasoline, ginger root and bat poop. Compare with English words for smells. They are generally compounds with *-smelling* and specific sources of smell (e.g., *gasoline, ginger root* or, why not, *bat poop*) but we don't have category words to describe the smells themselves. We do have some words that are used to describe smells at a more abstract level, such as *sweet, spicy, heavy* or *light.*

However, my own research on such abstract odor words has shown that there is little consensus on which odors are heavy and light. The same smell can be described by some individuals as *light* while others describe it as *heavy*. Majid argues that the poor ability to recognize and describe smells is a consequence of not speaking the right language. However, the results of worldwide research studies indicate that it is extremely rare to describe smells with broad, abstract words as Thai hunter-gatherers do. Only a few, usually very small cultures are characterized by such a developed olfactory language. Other cultures are, as far as we know, similar to us; they have difficulty naming smells and they use the names of specific physical sources of smell to describe them. Therefore, it may be that the olfactory brain is not adapted to generate linguistic descriptions. It takes unusual circumstances for us to overcome the brain's limitations and to successfully name smells when we don't have clues from the environment.

Although I believe that our brains have a limited capacity for naming smells, I also find it appealing to think that changing our environment can unleash our sense of smell. Researchers who have studied hunter-gatherers, such as Agnieszka Sorokowska of the University of Wroclaw in Poland, suggest that the absence of exhaust fumes and pollution makes rainforest inhabitants super-smellers. Given the sensitivity of our own sense of smell, we can only imagine what it would feel like if our noses were completely unaffected by exhaust fumes, and if we had been taught since childhood to put all of nature's smells into words. Perhaps we have additional nasal superpowers just waiting to be discovered. In the next chapter, I will show how our powerful brains guide our noses. The brain's influence on our sense of smell is usually beneficial, but in some cases it can make life unbearable.

CHAPTER 5

THE INTELLIGENCE OF THE NOSE

ONE OF THE most spectacular odor demonstrations was performed in 1899 by University of Wyoming chemistry professor Edwin Emery Slosson in front of his students. With a grave face, Slosson poured a clear liquid over a pile of cotton and told his students that a "strong and peculiar" but "not too disagreeable" odor would soon spread through the classroom, from the front rows to the back. He then used a timer hidden in his pocket to gauge their reactions. After just fifteen seconds, the students began to shift in their seats. More and more students raised their hands to show that they could smell it, first those in the front and then those in the back. After just one minute, the experiment had to be stopped—the smell was too intrusive for some students.

The reason why Slosson's experiment went down in history, as you may have already guessed, was that it was not a chemistry experiment at all, but it was of a psychological nature. The liquid was in fact distilled water. The smell was entirely imaginary.

It wasn't until 1972 that Slosson's experiment got a worthy follow-up. Trygg Engen, a psychology researcher at Brown University, conducted a series of experiments in which participants were asked to smell liquids, both mildly odorous and odorless, and judge whether or not they had an odor. Engen influenced participants' expectations by coloring some of the liquids with an odorless yellow dye. The color had strong effects on the responses he received. It made participants imagine that the liquid had a faint odor. One participant got 71 percent of the trials wrong when the odorless liquid had a distinct color, but only 29 percent when it was uncolored. Although the participants were told the correct answer after each trial, and could win some money if they answered correctly, they continued to frequently answer incorrectly: that the yellow, odorless liquid had an odor. It was, Engen wrote, as if an unconscious association affects our sense of smell. Thus, even after hundreds of trials, with both feedback and financial incentives, favorable circumstances that never occur in real life outside the research lab, it was still almost impossible to imagine that the bright yellow liquid was odorless.

It's not just olfactory judgments that can be affected by psychology researchers' tricks. Our brains' reactions to smells are shaped by what we think we are smelling. In one famous experiment, participants were placed in a brain-imaging scanner while a pungent smell was presented to them. Half of the participants were told that the smell came from a cheese. The other half were told that the smell came from sweat. The researchers were particularly interested in the area of the brain called the orbitofrontal cortex, located at the base of the frontal lobe, just above the eyes. This area is part of the brain's reward system, and it also connects impressions from smell, taste and other senses. The results showed that the smell produced completely different brain responses depending on how it was presented. When the smell was described as coming from cheese, the orbitofrontal cortex reacted strongly—the smell created a reward signal in the brain. However, when the smell was described as the smell of sweat, activity in the area was greatly reduced. Same smell, different brain reactions. The sense of smell is shaped by circumstances.

Some might say that such results show that the sense of smell is not reliable, but that it is a gullible, primitive and perhaps even a bit stupid sense. I disagree, and in this chapter I will explain why. I believe that we humans have an "intelligence of the nose," the ability to not only passively monitor the smells and odors in our environment, but also to weigh these impressions against knowledge, expectations and other circumstances.

Now consider the following situation: You are visiting a friend for the first time. You are let in and step into the hallway. There is a small smelling ceremony, which is over in an instant and barely registers in your mind. You sniff and take in the unique smell of the new apartment. For a short while it washes over your consciousness; you are surprised, wondering what it is that gives the apartment such a smell. The brain's olfactory processing spreads over ever larger parts of the brain via its interconnected nerve fibers. Sometimes you can find the answer to the question of what the smell is coming from. Freshly baked bread or coffee are familiar smells. But most of the time, the smell is unfamiliar, an unknown mixture that comes from the furnishings, the clothes, the floors and the people who use them. After a few seconds, you forget the whole thing and move on to thinking about the visual impressions, the furniture and layout of the apartment, and the conversation with your friend. The moment of smell is over.

How does this spontaneous olfactory ceremony take over our consciousness? To understand this brief moment, we can use scientific methods to study how the brain works and how it reacts to the smells we encounter. If we could look into the brain, it would resemble a chaotic urban environment. Imagine traffic flowing in every conceivable direction, people going in and out of buildings, eating food, talking to each other. The brain's activity is staggering. The methods for understanding the brain are not yet close to capturing all this activity. But every day new advances are being made.

So-called magnetic resonance images and electrical measurements of the brain's activity have given us a coarse-grained picture of how it works. But from this helicopter perspective, a picture of a well-oiled machine emerges. Individual areas have specialized tasks. Smells are processed in some areas, visual impressions in others. The different parts are linked to each other by neural pathways, like neighborhoods connected by traffic routes. Some areas seem not to be responsible for sensory impressions at all, but have more general tasks, such as storing impressions in our memory bank, or connecting different impressions—a face, a voice and a name—to each other. Other areas have even more abstract tasks. They have a kind of coordinating function, much like a manager of a football team. They make sure that certain players cooperate with each other, and can substitute specific players or the whole team if necessary. In the brain, individual areas rarely work alone but rather in teams, interconnected networks that can span the whole brain. When we need to focus our attention to solve a problem, the most appropriate areas come together and communication between them increases. When the problem is solved and we let our minds drift wherever they want, we replace the hardworking team members who are now resting. But this does not mean that brain activity decreases at rest. Instead, new teams enter the scene. These are the brain's cleaning and transportation workers, units that work when others are asleep. Thanks to them, life can go on as usual the next day, but we rarely think about their importance. My research team tries to understand how smells affect the different networks in the brain, and the results show that the influence of smells extends far beyond the "olfactory brain" and spans many areas of the brain.

Remember how the sense of smell works in disguise, and that what we call tastes are actually smells that come to the nasal cavity from the mouth, via the back door of the throat when we breathe out through the nose? The smells that come from inside the mouth are also one of the strongest pieces of evidence that the experience of smell is actually as much about your brain's interpretations as it is about the molecules you smell. The same smell can be perceived differently depending on whether the olfactory stimulus enters through the nose or from inside the mouth via the pharynx. Dana Small presented the same odors through the nose and through the mouth when experimental participants lay inside a brain-imaging scanner. The results showed that the brain was activated differently by the two stimuli, even though the same smells were presented. However, brain activity only provides indirect clues about how the smells are experienced—to really pinpoint the difference, participants need to be asked about their experiences. Together with researchers Curtis Luckett and Bob Pellegrino at the University of Tennessee, my colleague Thomas Hörberg and I designed a study that would provide a clear answer.

Bob and Curtis developed a method to distinguish the two odor stimuli. From a jar with a straw, participants "sucked in" odors that led to oral odor perception, but for the nasal odor stimulation, participants instead blew into the tube, releasing the odor under their nostrils. Thomas and I selected everyday smells that differed in both pleasantness and edibility, such as garlic, rose, banana and gasoline. We asked participants to guess the source of the odor and describe the odor experiences, and compared the results of the two odor methods. The result was the strongest evidence to date that the same smells can indeed be experienced differently—that the brain has two different "smell settings." When smells come in through the nose, the brain is tuned to explain the origin of the smell. The smells are then described with words linked to visual impressions, such as *colorful*. Participants also used more concrete words, such as *detergent*. When smells come to us from outside in this way, the brain wants to understand *what it is that smells*. However, when the smells come from *within*, from the mouth and throat, the brain seems to have a little more freedom, opening the door to personal associations, feelings and more abstract interpretations. The smells were then described using words associated with sensory impressions and internal states, such as *tickling*. The different ways of smelling stimulate the brain in different ways. Maybe that's why a wine taster is not satisfied with just putting his nose in the glass, but wants to let the wine wash around in his mouth before making a final judgment.

The sense of smell often seems to be in the background, in the shadow of consciousness, while what we see with our eyes is at the center of our attention. Psychology researchers therefore speak of "visual dominance" in our brains. It's an idea that has long been a feature of our culture. We still live in the "age of sight," and ever since the seventeenth century, exploration of the other senses has been secondary. Perhaps it is not surprising that smells are often forgotten. But sometimes new or unexpected smells appear and demand our attention. Sometimes it even feels as if a smell is taking over our entire thought process—there is simply no escaping certain smells. How does this work? Together with my colleagues, I have been trying to understand what makes the sense of smell different, and what happens when a smell attracts our undivided attention. Together with my colleague Thomas Hörberg, I created an experiment in which participants used keystrokes to classify fruits and flowers. Participants saw combinations of images and smells, such as a lemon with a lemon smell or a lilac flower with a lilac smell. Each item was followed by a tone that told participants whether they should classify the image or the smell. When the images and smells matched, the task was easy, and participants could quickly judge whether it was a flower or a fruit they had experienced. However, in half of the trials we made it difficult for participants because we made the items "incongruent," i.e., a flower smell was presented with a fruit picture, or a fruit smell with a flower picture. We were interested in how the incongruent objects distracted participants when making odor judgments and image judgments. What we saw surprised us.

Contrary to our expectations, and contrary to forty years of psychological research on visual dominance, it was not the images that distracted the participants the most. Instead, what was most distracting were the incongruent smells. Categorizing images when they coincided with incongruent smells was particularly difficult. On such trials, the participants were particularly slow. Their brain activity, which we measured in the meantime, increased, but was also delayed. This was a clear sign that distracting smells made the task particularly difficult. But it raised another question: Why are unexpected smells so noticeable?

To understand how important anticipation and surprise are for the sense of smell, we should think about one of our most basic living conditions—the fact that we need energy to survive. Just like other animals, we have to search for our energy sources. This presents us with two major challenges: we must find nutritious food, and we must avoid being poisoned. The sense of smell seems to be perfectly designed to help us perform these two tasks. And the approach is simple but ingenious. If you were to sniff and try to smell your surroundings right now, you probably wouldn't feel anything special. Your nose has become accustomed to the smell of the place. This is because the receptors of the sense of smell get tired quickly. Within a minute, the sensitivity to smell has been fatigued so that it now takes a much stronger dose of the smell to create an olfactory signal to the brain. This mechanism, known as desensitization, has been well known for over a hundred years. Back in 1935, American neurosurgeon Charles Elsberg conducted experiments with perhaps the most popular smell of all: coffee. Elsberg pumped the smell of coffee into participants' nostrils and carefully measured how strong and prolonged this stimulation needed to be to result in numbing. After just three seconds of exposure to a strong coffee smell, the participants became numb to the smell, and it took six minutes for their sensitivity to be fully restored.

Desensitization plays an important psychological role by enhancing the perception of *changes* in the olfactory environment. Desensitization makes us ignore background odors—unless they change! This is because our receptors are specialized, and we only experience the fatigue effect when we are exposed to the same smell for a prolonged time. When the smell changes, new, well-rested receptors are called in from the bench and our attention is drawn to the new smell. The automatic redirection of attention causes the brain to allocate more thinking power to the smell to assess whether it is the dangerous smell of fire smoke we are smelling, or whether it is the appetizing steak from the restaurant on the corner. When a new, faint smell approaches it is not drowned out by all the background smells, because we have become numb to them. This helps us detect the sources of energy as well as the dangers in the environment.

The stronger the odors and the longer or more often we are surrounded by them, the more time is required for recovery. While desensitization appears to be a simple, primitive mechanism, Elsberg's old coffee experiments showed that nasal desensitization was more sophisticated than people might think. In one experiment, he pumped the coffee smell into only one of the nostrils. Each nostril carries smells to the nasal cavity, which consists of two vessels separated by a partition, so that smells travel via parallel paths up to the receptor neurons and from there to central parts of the brain. If desensitization was simply due to the fatigue of the receptors in the olfactory mucosa, the unused nostril would retain its sensitivity, whether or not its neighbor in the adjacent nasal cavity was dulled. But the results pointed in a completely different direction. Sensitivity was lost to some extent even in the nostril that had not been stimulated. This led Elsberg to understand that desensitization does not only take place in the nose, but also *in the brain*. The loss of smell in the nose is thus partly psychological in nature. The brain seems to simply forget about smells after a period of exposure. Psychological experiments have shown that if our attention is directed away from smells, we become numb to them more quickly. Our unconscious thought processes affect how much we become numb. Even when we have become numb—for example, on a visit to the perfume shop—we can still smell strong smells, but in a weaker form, and only if we concentrate. Unfortunately, there are no smells that increase sensitivity to other smells, although perfume sellers sometimes allow customers to smell coffee beans to "clear their noses."

Just as the brain deprioritizes smells to which it has become accustomed, its reward system is dampened when we have just consumed a fragrant food. This is a sign that it is time to move on to something else. This was investigated in a research experiment that sounds almost too good to be true, at least at first glance. The researchers asked their participants to eat pieces of chocolate while their brain activity was imaged. But first, the researchers themselves indulged in fifteen different types of chocolate—this pleasurable experiment led them to choose the Lindt brand of chocolate, which they found most delicious. Participants had to start with a piece of chocolate, let it melt in their mouths (chewing should be avoided if you're in a brain scanner because your head should be completely still) and rate how good it tasted, or, actually, smelled, because the chocolate "taste" is really all about its aromas. The participants gave the first chocolate almost the highest score, 9 points out of 10. The next piece was tried in the same way, but now the same chocolate scored only 8 out of 10. So the experiment continued, piece by piece, with less and less enjoyment. After only seven pieces, the chocolate received a *negative score*. Since the participants' brain activity had been measured during this time, the researchers could see how certain brain areas decreased their activity as the pleasure decreased. This suggested that these regions were chemical pleasure centers. Other areas had the opposite characteristics: they increased their activity as the chocolate became more unpleasant, indicating that these areas signal the body's satiety, or perhaps even nausea. Sometimes the different areas were in close proximity to each other, and it is conceivable that they communicate frequently with each other. Together, these areas can give us clear signals that the chocolate is now not very appealing, and that we should move on to new appetizing smells and tastes.

Research on desensitization is important for understanding the sense of smell. Unfortunately, it has also contributed to the widespread belief that the sense of smell is passive and only reacts to changes. According to some scientists, the sense of smell only reacts when we smell new odors, which would mean that it is completely disconnected from the brain's knowledge and thought processes. But this theory leads us to underestimate how smart the sense of smell really is. There is strong evidence that it can make use of the brain's stored knowledge, even if we have difficulty expressing this knowledge in words. The sense of smell is "silent," but it can use other sensory input to figure out what we are smelling. The brain uses what we see and hear to predict what the next sniff will smell like. It is thanks to this tacit olfactory knowledge that we rarely have to think about the smells around us—except when we are struck by an unexpected smell.

But can the sense of smell really make predictions about the future? A particularly interesting example is people who are severely bothered by perfumes and other smells around them, common smells that are completely unproblematic for most people. The reaction has been given many different names, but is often referred to simply as chemical sensitivity, and it is reported by about 10 percent of the adult population, i.e., 26 million US adults. For those who are sensitive to odors, a visit to the cinema or a bus ride can be unbearable. The smells from fellow passengers, exhaust fumes and perfumes cause increasing nausea, headaches and dizziness. The sufferer has to move away from the source of the smell in order to relax again. These people suffer, sometimes severely, from their condition and avoid environments where strong smells may be present. This is why you may see signs in hospitals and other workplaces urging you to avoid wearing perfumes and other strong-smelling products. The fact that we experience smells so differently is a constant source of unhappiness for those who are particularly sensitive. Your favorite perfume, which for you is enjoyable, subtle and tasteful, can be a real nuisance for these people, comparable to a scream or music at high volume.

To better understand the phenomenon of chemical sensitivity, let's return to a figure we met earlier: Marcel Proust. He had suffered from asthma since childhood, and was not afraid to appear fragile. Physical weakness and sensitivity were a sign of intelligence, according to the standards of the time. Historian Karin Johannisson paid a lot of attention to the gray area among illness, psychology and the changing historical gender roles that make men like Proust so odd today. Johannisson writes about Proust in the book *Melancholic Rooms* that as the best man at his brother's wedding, he appeared in "triple coats, several pairs of gloves, and with his chest and neck wrapped in several layers of wool. He had been ill for several months, he explained, and was about to fall ill again." Proust was highly sensitive to virtually every sensory impression he encountered. In his apartment at 102 Boulevard Haussmann in Paris, the walls were lined with cork to isolate him from disturbing sounds. The curtains were constantly drawn to prevent any light or noise from entering.

Of particular interest to this story, which is after all about the sense of smell, is that Proust could not stand many smells. Johannisson says that Proust disliked almost all food, and the smell of food made him nauseous. No food was therefore allowed to be cooked in his apartment: *The only smell allowed is that of his smokes. To light them, a wax candle must burn day and night. He can't stand the smell of sulfur from the matches, nor the sound of them being struck on the floor.* So what did Marcel Proust smoke in his apartment? Opium, among other things, a drug closely related to morphine that is an analgesic and provides a numbing high. One of the hallmarks of the opium rush is *synaesthesia,* a confusion of sensory experiences. The overflowing wealth of sensory impressions that Proust conveyed to his readers was certainly influenced by the mind-expanding substances he ingested in his chamber.

People suffering from chemical sensitivity are often described as "the family bloodhound," who can smell odors that no one else notices. The reason, many believe, is that they have particularly well-developed and reactive receptors in their nose. But researchers have not found strong support for this theory. Chemically sensitive people are not only sensitive to smells, but they often react strongly also to noises, a phenomenon that cannot be explained by smell receptors. Chemically sensitive people do not have particularly sensitive noses, as their thresholds for detecting faint odors are in the normal range. Instead of sensitive noses, I believe that they have sensitive brains. Research experiments show how their reactions may be elicited. When a chemically sensitive person sits in an odorless chamber that they know eventually will slowly fill with an increasingly strong smell, their brains react immediately, their stress level increases and they start thinking about the smell they will be exposed to. Their ability to solve cognitive tasks on a computer screen is impaired as all attention is drawn to the smell—even though it hasn't even begun to permeate the chamber's vents. Once the smell arrives, the chemically sensitive participants feel as if it is growing stronger and stronger. The average-sensitivity participants soon forget about the smell and can focus on their other tasks. However, the brain's usual numbing process, which usually causes us to forget about the smells around us, has backfired on the chemically sensitive, and instead begins to amplify the odor signal until it becomes unbearable. The chemically sensitive find everyday smells too strong, but even when the researcher presents an empty, odorless bottle, the chemically sensitive often find an odor there. They may have unusually strong reactions to smells, but they illustrate a more general point about smelling: our reactions to smells are never just about molecules entering our nose. The sense of smell doesn't just react to changes in the environment—our brains create expectations that can sometimes be so vivid that they are hard to distinguish from real-life smells. If we are prone to feelings of anxiety related to smells, our brains make us worry about what smells to expect when we get on the bus or enter the movie theater. The sense of smell is not *reactive* but *proactive*.

Chemical sensitivity is a terrible affliction for many people, and we need more research on how to help them. However, sufferers sometimes dismiss psychological research—they say that their problems are not psychological at all but due to a sensitive nose or allergic reaction, and that only doctors can understand their problems. But the truth is that the best research on chemical sensitivity is often done by psychology researchers, who can understand and accurately measure how expectations, anxieties and stress affect our olfactory experiences. It's not a matter of "imagination." It's about the brain making predictions and creating stress responses that become overwhelming. Smells are rarely dangerous, but that's little comfort. In fact, the same goes for elevators, airplanes and open town squares. They're not dangerous in themselves, but phobias can nevertheless shake some of us to the core. Phobias are very real—they involve both our psychology and our brains—and they are effectively cured using cognitive behavioral therapy. Such treatments are now being tested to help those suffering from chemical sensitivities too.

So the sense of smell often draws on other parts of the brain. Therefore, the olfactory brain's activity increases when we see or hear something that reminds us of an odor, such as the sound of a simmering coffee maker, the image of a bouquet of flowers or the word *cinnamon.* Our research shows that the sense of smell, more so than the sense of sight, is guided by verbal cues. In one psychological experiment, participants are given cues about an upcoming smell or image. Sometimes it matches the cue (such as when the word *rose* is followed by a matching rose smell, or the image of a rose), sometimes not. We measure how much faster participants respond when the smell or image matches the previous cue. When the cue was a match, the speed increases particularly in the sense of smell—a sign that the brain has prepared for that particular smell or image. The olfactory brain thus seems to absorb the cues coming from other parts of the brain, creating an expectation of what will come next.

How do the olfactory brain and other brain regions work together? My colleagues at Northwestern University in Chicago have been studying patients who have electrodes inserted into their brain tissue. This is a necessary procedure for epilepsy patients before brain surgery, and the patients have local anesthesia but are awake during the procedure. This has allowed us to understand how the olfactory brain reacts to verbal cues about an upcoming smell. We see that the olfactory brain starts to prepare for the upcoming odor even before it is presented. A synchronized oscillation of electrical activity suggests that the olfactory brain and the auditory brain are communicating with each other. And when this preparation does not take place—when the olfactory and auditory brains do not oscillate in tandem—the patient also fails to determine whether the cue matched the smell. The results have led to a new explanation for how we manage to recognize smells. It involves several different areas of the brain dancing to the beat, linking together smell, knowledge, emotions and associations. Indeed, the olfactory brain is particularly influenced by cues from other senses, as it has a particularly strong interaction with other brain regions. The insight that the sense of smell is shaped by expectations and environmental cues is of great practical importance. It can help us develop new methods to cure people who have an insensitive—or hypersensitive—sense of smell.

The sense of smell does not act on its own, but is smarter than that—it takes in all the cues in the environment and assesses them using all our accumulated knowledge. Our brains not only react to smells, but also make predictions that determine what we think we smell and how we react to the smells. The sense of smell is expectant and interpretive. Many companies want to use smell to influence our emotions and behavior, and if they want to be successful, they need to understand the special intelligence of the nose. The next chapter is about the olfactory industry.

CHAPTER 6

FOLLOW YOUR NOSE

ONE DAY I received an email from an unusual sender—a consultant from one of the leading real estate companies in Sweden contacted me and wanted to meet. I met with the consultant and learned that the company was trying to use scents in office environments to create value for their customers. Their goal was that the scents would lead to increased productivity, improved well-being and make the company's office better than its competitors, thus increasing profits. What scents make people more focused, happier, more social and more likely to stay at work longer? I did my very best to explain what the research says, but the links among smells, emotions and behaviors are much more elusive than you might think.

Fragrances are a mega-industry. According to industry reports, it is valued at around USD 50 billion per year. The spice industry is valued at 20 billion dollars. If the sense of smell was not important to us, we would hardly spend so much on these products. The wine industry, whose success is largely due to the complex aromatic notes of wine, is valued at 300 billion dollars a year. But fragrances are not limited to the domain of food and drinks. In fact, most of the products you encounter in everyday life have been made with fragrances in mind. Fragrances are often added to products and materials to mask bad smells. Bad smells are a common problem, especially in the production of textiles and plastics. If all the plastic products we use retained their "natural" smell, they would in many cases be avoided by customers. Smells are also used to create positive associations with products. The specific smell of "new car," for example, can be bought as a spray and makes old cars feel fresh and new—one of many tricks used by car salesmen. Something about these smells must be important to us, even if we don't always understand it ourselves.

The key to understanding how smells affect our behavior is the emotions that smells evoke. Emotions are perhaps the most prominent feature of smells, and many researchers have tried to find a "molecular recipe" to explain our likes and dislikes. By analyzing the molecular structures of different smells and examining how people evaluate smells on an emotional level, they believe we can gain a deeper understanding of how smells create emotions. The researchers use advanced statistical tools that can find links between different chemical properties of the odor molecules and the emotions that the odors evoke.

The new science of odor is both promising and concerning. Can odors be applied to influence people's behavior, so that we work more efficiently or buy certain products? I am skeptical of such predictions. I believe that smells are perceived differently depending on our individual experiences and associations. But an odor that evokes a positive emotional response in most people is of course preferable in public environments to one that evokes distaste—that much is probably obvious. Researchers who find that certain molecular properties are associated with positive emotions may sound impressive, and they may impress companies that want to gain an advantage over their competitors—*see how my advanced algorithm can select exactly the fragrance molecules that create positive emotions in the customer!* But this research doesn't tell us how the connection is made, whether the emotional associations of these molecules are innate or learned.

It could be that most people simply like the molecules that they have already learned to like (such as fruits and flowers), and unfamiliar smells are liked to the extent that they resemble those previously learned positive smells. The high-tech algorithm may just be a complicated way of telling us what we already know: that people tend to like some smells but not others. And the best way to find out which smells customers like is still the simplest—ask them!

A lot of research has been done to find out what smells people like. And in many cases we agree with each other. An Israeli research study identified ten smells that participants found particularly pleasant:

Lime
Grapefruit
Bergamot
Orange
Peppermint
Freesia (flower smell)
Banana
Cassia (cinnamon smell)
Mimosa tree
Fir tree

Most of the smells on the list you would probably recognize or find familiar (cassia is a cinnamon-like smell, and bergamot is a citrus smell found in many perfumes). There is a strong correlation between how familiar a smell is and how pleasant we find it, so if you want to create a smelly office environment or scent a shopping mall, you should choose smells that are familiar to your target audience. But you need to think about more than just the smell. In a noisy, dirty or otherwise messy environment, no background smell in the world will win the hearts of customers. On the contrary, the smell will take on the negative qualities of the other senses and may end up being perceived as "cheap," "artificial" or even "nauseating."

In addition to being both malleable and teachable, the sense of smell has other peculiarities that cause problems for those who, like my consulting colleague, want to find a scent that makes people more focused at work. The individuality of the sense of smell means that some smells suit some people but not others; for example, men and women seem to have different smell preferences in some cases. At least if you believe the European research team that ranked a large number of smells according to how much the genders differed in their assessment. Starting with the nasty smells, it's clear that women really hate strong body odors. The smell of sweat and the smell of old socks were particularly disliked by women. The smell of cigarette butts, smoke, strong cheese and onions also caused more discomfort for women than for men. But a striking number of women were also negative about smells that are usually considered positive, such as coconut and lily of the valley. Men were on average more accepting of smells, and only a few smells were less popular with men than women. These smells included cucumber, honey, lavender, lilac and—perhaps unexpectedly—coffee!

Anyone who thinks they have found a suitable odor to scent a shop, café or office environment soon faces another problem. How strong should the smell be? It is not necessarily the case that an odor is equally pleasant at all levels of strength. On the contrary, many smells are assessed differently when they are weak compared to when they are strong, but as a rule, strong smells are perceived as less pleasant. Weak or medium smells give the best results. A few smells, such as citrus and jasmine, are exceptions—they can even become more pleasant as their strength increases. But even here there are reasons to be careful. Working in an office environment that smells strongly of jasmine risks turning a jasmine lover into a jasmine hater. Don't overestimate people's tolerance for an intrusive and persistent smell—and remember that the chemically sensitive are distracted by the mere thought of strong smells.

The consultant's question, whether fragrances could increase productivity, sparked my interest and made me dig deeper and deeper into a veritable sea of research reports. *Swamp* is perhaps a better description of the state of the research, because the picture actually became more and more clouded the deeper my knowledge grew. To begin with, there is undoubtedly strong evidence that smells can increase people's alertness and heart rate. The most well-known example of this are so-called smelling salts, which throughout history have been put under the noses of fainting ladies, battered hockey players and boxers on the verge of collapse. This smelling salt usually contains ammonia. In addition to smell, ammonia stimulates the trigeminal nerve,* which transmits tingling, burning and cooling sensations to the eyes, nose and mouth. Ammonia has a strong activating effect on our bodies. It raises the heart rate and blood pressure, which leads to blood being pumped to the brain and increases the level of alertness (similar effects can be obtained by drinking a cup of coffee, but coffee is harder to force on a semiconscious pugilist between rounds).

The ammonia-based smelling salts have strong effects, and in fact, smelling salts are now banned in the boxing world, as it is considered too dangerous to temporarily awaken a boxer who should be throwing in the towel. And the smell of ammonia is not something an employer should expose their employees to. It simply smells awful, so a lot of effort is put into keeping ammonia odors out of our noses. One of the most well-known odor spreaders is the pig, whose feces and urine produce a strong ammonia smell. Long-term exposure to ammonia can actually pose a health risk for both the pig and the

---

* The trigeminal nerve is one of the human body's twelve cranial nerves leading to the brain. It has branches in the mucous membranes of the eyes, nose and mouth and transmits irritating, tingling, cooling and burning sensations from these areas, when we eat spicy food, for example.

farmer who stays in the pigsty for long periods. This is why a limit value
of ammonia has been established at 10 ppm, or parts per million. There
must not be higher concentrations of ammonia in the air inside the pigsty, as this can damage the respiratory system.* You can get used to most
smells in reasonable doses, and I have fond childhood memories of the
smell of manure on my grandparents' farm. But I am sure that even the
most dedicated pig farmers would prefer pigs that smelled of citrus and
jasmine rather than ammonia, if such pigs were available. Ammonia is
simply not something you want in the workplace. However, there are
other ways to achieve a stimulating odor effect. Commercially available
"odor salts" use more pleasant smells. For example, the pungent, cooling or spicy aromas that come from peppermint, wintergreen oil and
eucalyptus. But even these smells risk irritating the airways after a short
time. So it's better to keep these fresh smells in a small bag on your desk
than to pump them into the central ventilation system.

Increased alertness could be an effective selling point, but as I
have described above, the most stimulating smells are also those that
are often the most difficult to inhale for a long time. But smells can
also play on other emotional strings. The stimulating effect is not the
only positive effect that can be achieved with the help of smells.

Scent marketing is a small but growing marketing niche with
an annual turnover of USD 350 million. These companies develop
specific "scent concepts" for their customers, which are usually hotels
and retail chains. These are companies that want to enhance the experience of being on their premises and make customers, and staff,
emotionally attached to the brand. One scent marketing company
claims on its website that 35 percent of Fortune 500 companies use
fragrances as part of their brand.

---

* Minimizing ammonia odor from pigsties has become something of a science:
  experts calculate the ventilation system, ceiling height and airflow of the sty to
  optimize the exit of ammonia odor, but also behavioral science methods are used
  to teach the pig to do its business in the right place.

A simple example of scent marketing is the American café chain Cinnabon, which is famous for its cinnamon buns. When you are in the vicinity of a Cinnabon café, you can smell the familiar cinnamon scent from afar. This is no coincidence. Cinnabon cafés are always placed inside malls, never outside, so that the smell does not blow away in the wind but spreads to as many customers as possible. Cinnabon has also adapted its ventilation system to ensure that as much cinnamon smell as possible reaches the customers from the bakery. As if that weren't enough, the chain also allegedly uses an artificial cinnamon scent that is dispersed in the air to further reinforce the olfactory message. Products containing competing scents, such as onion or garlic, have been removed from the Cinnabon menu so as not to interfere with the olfactory experience. However, for those of my fellow Swedes who want to try this rather extreme form of scent marketing (and the American-style icing-soaked cinnamon buns), I have some bad news. The chain's only store in Sweden, located in the Kista Galleria outside Stockholm, closed due to the COVID-19 pandemic in November 2020. Cinnabon's future in Sweden may be uncertain, but fortunately there is not exactly a shortage of cinnamon buns in Sweden, and many other cafés and grocery stores now display their freshly baked cinnamon buns near the entrance to their stores so that no one misses their scent.

Using cinnamon scent to attract customers to a café chain is simple and clever. But using scents to build brands for hotel chains, airlines or clothing brands is a much more complicated task. In developing a suitable scent profile, scent strategists use a mixture of science and artistic freedom. Typically, they first try to understand the customers' existing brand—what values and associations people have with the brand, and what values, emotions and associations they want to create or reinforce. Then it may be time for a fragrance strategist to take action. I will return to such an example shortly.

There are many smells that are said to have special inherent properties to affect us psychologically and health-wise. It is not uncommon for people to want to diffuse the smell of oils from flowers and other plants in their homes or apply them to their bodies in order to improve their health and well-being. This is called aromatherapy, and the US aromatherapy market alone is said to be worth around USD 2 billion annually. Claims about health benefits are used to sell the aromatherapy fragrances to customers, and in the US supermarket chain Target, for example, you will often find an aromatherapy shelf with various fragrant oils. Online and in the marketing of fragrance companies, descriptions of the psychological effects that different smells are believed to have are offered: lemon increases concentration and has positive health effects, rosemary promotes memory and counteracts fatigue, lavender counteracts stress and depression, jasmine promotes independence and vitality, cinnamon improves concentration, peppermint promotes clarity and gives energy. So a lemon-rosemary-lavender-jasmine-cinnamon-peppermint blend under the nose would be the recipe for a happy and productive life? No, it's not quite that simple.

Aromatherapy sits in the gray area between science-based knowledge and pure humbug. It's actually difficult to pinpoint exactly what aromatherapy is, as it's not just about inhaling a smell—sometimes eating capsules containing these natural oils is also considered aromatherapy—but I'll focus on the smells here. Aromatherapy is classified by the US Food and Drug Administration as an "alternative medicine" practice, as the medical effects of aromatherapy are not considered to be scientifically proven. Nevertheless, one sees claims about the health effects of odors, often with references to more or less scientific studies. But this does not mean that the effect of aromatherapy has been "proven" to be true. Scientific references can be misused. In aromatherapy research, it is common for researchers to ask participants to perform various tasks while smelling lavender, for example. In some studies, participants are given a massage, with or without the smell of lavender. Then the participants rate how they feel, often using numbers on a scale: *How relaxed do you feel right now?* Another group is in the same situation, but without the background smell. By comparing the groups' ratings, feelings or performance, researchers sometimes find differences that they think are meaningful. Perhaps the lavender group judges that they feel more relaxed than the comparison group? Using statistical tools, the researcher calculates the risk that the results could be a coincidence, and if this risk is deemed low enough, the researcher concludes that lavender really does help people to be more relaxed. In this way, a whole industry has grown up around aromatherapy. Lavender is one of the most common smells in aromatherapy research. But anyone with a deeper understanding of how research works can easily become suspicious of many of these studies and start asking critical questions.

Let's start with the comparison itself: the smell of lavender is relaxing compared to no smell. Even if lavender's scent is indeed more relaxing than having no background odor at all, it may not be a particularly remarkable result. It doesn't say anything about whether lavender has special olfactory properties, because then it would be more relaxing than other smells. The research does not show this. Anyone whose business idea is to sell lavender scents to stressed-out customers who want to relax should be able to show that lavender is *more relaxing than other scents*. There must be a reason why the customer should buy the jar of lavender instead of sniffing the jar of cinnamon in the spice cabinet. But already here, at this basic level, many aromatherapy studies fall short.

In more well-controlled studies, different smells are compared with each other. This raises the bar for finding meaningful results. The comparison odor should be similar to the experimental odor; for example, it should be equally strong and equally pleasant, but it should lack the property that researchers assume produces a potential psychological effect. Cinnamon could be such a comparison smell. If the results still suggest that the lavender group feels more relaxed than those who smelled cinnamon, then one might begin to suspect that lavender really does have a relaxing effect. But we're not talking about strong evidence yet. The relaxing effect may actually be due to the participants *recognizing* the smell as an aromatherapy smell, which makes them feel safe. This is what I believe is the reason why some aromatherapy smells are perceived as relaxing. In fact, research has shown that if we are asked to smell a "chemical," we dislike it more and become tense, compared to if the same smell is presented as a "natural" or, even better, an "aromatherapeutic" smell. The molecules of the smell are not decisive, but our reactions are also influenced by the expectations of our brain as it guides our noses.

The cognitive perspective is thus important for understanding how aromatherapy works. The psychological effects of the lavender smell can be strong, perhaps even as strong as a drug, but the effects of the smell—unlike the effects of drugs—depend largely on our individual memories, associations and expectations. For those who have been taught that lavender is relaxing, this may well be the case. But for those who believe that the smell of lavender is a health hazard, it will instead cause stress. Experiments have been carried out to prove just that: half of the participants were led to believe that the smell of lavender could be harmful, the other half that it had a therapeutic effect. The same lavender scent, when associated with these different expectations, had dramatically different effects on the participants' perceived stress, but also on their blood pressure and heart rate. Thus, it is not the odor molecules that determine our reactions, but we ourselves. The fact that people are not uniform robots, but individual persons with unique memories and emotional lives, is what makes olfactory marketing and aromatherapy as much an art form as a science. In some cases, the marketing even resembles humbug. The emotional messages that smells may convey are inherently personal, as they are colored by our individual experiences.

The olfactory strategist who wants to try to shape an olfactory experience that can be shared by most people must combine smell with a strong and positive message—a brand, value or feeling. Over time, customers can then learn to embrace the smell and the positive message it represents.

A Swedish hotel that wanted to use smells in its branding hired an olfactory branding company to develop a completely new "signature smell" for the hotel. The smell was given a prominent role in the company's entire brand. Visitors to the hotel are greeted by the smell right at the entrance. The hotel's shower soaps and shampoo bottles have been scented in the same way, and these products are of course available for sale at the reception desk. When customers come home from their relaxing weekend at the hotel, the smell of the shower soap and shampoo will help them maintain the positive memories—hopefully until it's time to book their next hotel stay.

Olfactory strategists have a variety of smells to work with to compose their signature smells. They present a pictorial story to the customer about exactly what emotions and associations the smells are expected to evoke, and how they fit with that particular customer's brand. But does it work? It is difficult to evaluate whether this is a strategy that increases profits for the company. Many factors come into play because the sense of smell engages the human brain in such a complex way. What are customers' previous associations with the signature smell's notes? Will they form new, positive olfactory memories of their stay? Or will the constant presence of the smell become intrusive and unpleasant? Although there are no easy recipes for olfactory marketing, I believe it is here to stay, as a natural part of companies' efforts to communicate a personal identity that speaks to customers on an emotional level.

Aromatherapy will continue to exist as long as people feel positively affected by smells. The belief in the healing powers of smells has a history going back several thousand years, from ancient incense to medieval miasma theory. Good smells carry with them a cultural and historical legacy that leads our minds to health and well-being. The cognitive perspective on the sense of smell helps us understand why we are so strongly affected by the smells around us—they are shaped by emotional experiences and memories. But it also helps us realize why this impact is quite individual—our feelings and memories are often uniquely personal. In other words, the ability to control or manipulate people through smell is limited, and thankfully so. One particularly controversial olfactory phenomenon is pheromones, chemical substances that are said to influence our behavior in an unconscious way. In the next chapter, I will examine the concept of pheromones: Which claims are true and which can be dismissed as myths?

CHAPTER 7

SMELLS AND INSTINCTS

S CENTS, ODORS AND smells are an inevitable part of our lives. Few sensory impressions are as effective in arousing our appetites and desires, but also our disgust, as smells. This ability has led to the sense of smell being described as a primitive and animal sense. There are many beliefs about the sense of smell and how it affects us. In the world of smell, our innermost, primitive drives and desires are given free rein. Nowhere are these desires more vividly described than in Patrick Süskind's novel *Perfume: The Story of a Murderer.*

The book quickly became an unprecedented success when it was published in 1985. The story revolves around the mysterious murderer Jean-Baptiste Grenouille, who is born in a fish market in eighteenth-century Paris and has a superhuman sense of smell. Grenouille never fits in in the stinking French capital. People shun him because he has no body odor at all. Grenouille's olfactory alienation grows over the years, as he pursues a career as a perfumer while cultivating secret plans for world domination. Using the body odors he steals from the teenage girls he murders, he mixes a perfume to control people's minds. Here's how he describes one of his victims, whom he watches from a distance: *Her sweat smelled as fresh as the sea breeze, the sebum in her hair as sweet as nut oil, her sex like a bouquet of water lilies, her*

skin like apricot blossoms . . . and all these components combined to make
a perfume so rich, so balanced, so enchanting. . . . This scent was the
higher principle by which the others must be ordered.

Süskind's story of the murderer Grenouille has sold over 20 million copies since it was first published. Its themes of social exclusion
and desire made it a favorite of 1990s alternative rock stars Marilyn
Manson and Kurt Cobain. The narrative has a seductive allure; its
vivid descriptions of smell reflect a fascination with how they can affect our emotions, thoughts and behaviors. Public interest in smell
has been particularly focused on *pheromones,* perhaps the most controversial topic in olfactory research.

The term *pheromone* was coined in 1959 by scientists Peter Karl-
son and Martin Luscher, from the Greek *pherein* ("transport") and
hormone ("drive"). The scientists were summarizing the new research
on how insects communicated with each other using chemical signals. The first pheromone was discovered by German scientist Adolf
Butenandt and his colleagues when they realized that female silk-
worms secreted the substance *bombykol* to attract males. A pheromone
is a chemical, hormone-like signal secreted by an individual that leads
to specific behaviors or biological responses in other individuals of the
same species. The big mystery is whether pheromones also exist in
humans. Can unconscious body odors control our behavior and who
we are attracted to? This question has occupied scientists since the
birth of the concept more than sixty years ago.

It has long been known that animals communicate by sniffing
each other to get clues about fertility and other traits. In the Lascaux
caves in southern France, a seventeen-thousand-year-old painting of
an ox appearing to smell a cow has been found, with the smell itself,
coming from the cow's backside, illustrated in color. Modern research
has confirmed that pheromones are common in the animal kingdom.
They are particularly common in insects, and they influence mating
behavior and navigation; for example, ants use pheromones to find

their way back to the stack, and wasps use them to guide each other to new colonies. Many mammals also use hormonal secretions, which are important for parent-child bonding. Dogs, cats and birds use pheromones to mark their territories. It stands to reason that most animals, including humans, are affected by pheromones.

Research on human pheromones is of great interest, but their existence has been difficult to prove. Controversial studies claim that smelling alleged pheromones causes specific reactions in an area of the brain that controls the body's hormone levels and is therefore important for our sexuality. In one study, researchers found that homosexual men and heterosexual women reacted strongly to the substance secreted by men, while the substance secreted by women caused the same reaction in lesbian women and straight men. The results were interpreted as indicating that homosexuality has a biological cause. However, these controversial studies had a limited number of participants and the results have not been confirmed by other researchers. This suggests that the results may have come about by chance. And given that the concentration of the alleged pheromone substance was more than one hundred times higher than that found in real-life body odors, it is unclear what practical significance the results actually have.

The strongest critic of pheromones is olfactory scientist Richard Doty. In his book *The Great Pheromone Myth,* he describes how the alleged pheromones of mammals are not as powerful and automatic as one might think. In fact, pheromones are highly influenced by learning and social circumstances. Thus, unlike insects, mammals do not seem to be controlled by the automatic influence of chemical signals, but have a greater degree of freedom and flexibility. The original meaning of the pheromone concept was that they control the receiver in an automatic way. Instead, mammals often need to train their responses to pheromones. A baby mouse learns to associate the smell of its mother with food and care, and so the smell becomes increasingly important. Unfortunately, pheromone researchers have too often ignored the cognitive perspective, and they have contributed to the popular, but in my opinion incorrect, notion that pheromones have a strong and unconscious influence on us. Doty is right about many things, but he goes unusually far in his criticism when he says that the pheromone concept is not applicable to mammals at all. Most researchers try to find a middle ground, saying that while the effects of mammalian pheromones can be affected by learning and other factors, the pheromone concept is still useful. They also point out that even some insect pheromones have context-dependent effects—not even insects turn into robots as soon as they detect a pheromone smell, but their behavior is influenced by whether it's an appropriate time for them to eat or mate, for example. However, when we acknowledge the cognitive perspective, that both mammalian and insect pheromones are context-dependent and influenced by learning, the original meaning of the pheromone concept, where effects were assumed to be automatic and predictable, also disappears to some extent. The line between what is a pheromone and what is simply an ordinary smell has started to blur.

However, individual breakthroughs have occurred relatively recently. A French research team led by Benoist Schaal and his PhD student Gérard Coureaud discovered in 2003 that female rabbits secrete a chemical substance, 2-methylbut-2-enal, which triggers a sucking reflex in their newborn babies. The survival value of this behavior for the young is obvious, as they have to suckle for nutrition immediately after birth. Coureaud has since dedicated much of his research career to deepening our knowledge of this pheromone. Together with colleagues, he has shown that the pheromone is extremely powerful in the rabbit pup, with a brief exposure to the odor molecule being enough to trigger the sucking reflex with 100 percent accuracy. The substance also seems to act as a door opener for other types of learning. The baby rabbit can remember other smells paired with the pheromone, smells it would otherwise have forgotten. Coureaud is a creative scientist who has managed to develop his early breakthrough finding into an entire research program on odor perception and memory. But when I met him in a restaurant in New York, he told me that he still sometimes felt a bit disappointed. Almost all mammalian brain research is still being done on rats and mice, and brain researchers have so far shown little interest in starting to work with rabbits, he told me. The mice and rats don't seem to have any pheromones similar to what he discovered in the rabbit. His pheromone is an amazing discovery, and is also perhaps the single greatest success of pheromone research in the mammalian world.

It may seem reasonable that humans would have pheromones, as most of us are actually very sensitive to smell. Like many other animals, our bodies emit odors, and these body odors tell us about our health, our diet, our age and our stress levels. Our body odors also change throughout our lives. This is most evident during puberty. The body's hormonal changes cause new substances to be released from the sweat glands in the hairy skin under the armpits and at the genitals. When these substances are broken down by skin bacteria, they release other substances that lead to our familiar body odors: *valeric acid,* which is similar to cheese, and *propanoic acid*, which is similar to vinegar. These smells are disgusting, and we spend a lot of energy and money trying to minimize and hide them. But pheromones, which are a type of odorant emitted from the body, are thought to have an attractive effect. How does this work? The body's secretions contain several odorants that are induced by hormonal processes. Men's sweat contains testosterone-like substances. The best known is *androstadienone* (AND), which is said to be secreted by men to attract heterosexual female partners, and in women there is *estratetraenol* (EST), an estrogen-related substance that is said to have a similar effect on heterosexual men. The idea that AND and EST are human pheromones has become well established.

The idea of body odor's power over us has taken root in popular culture. Pheromone perfumes have long been available online, for example, on Amazon.com. The company RawChemistry sells pheromones for both women and men. The female pheromone perfume contains estratetraenol and *copulins* (compounds found in vaginal secretions that are presumed to be pheromones). It is claimed to have a proven ability to attract men, as the chemicals attach to their "sexual receptors." The male pheromone perfume contains a mixture of different substances derived from male sex hormones, such as androstadienone. Both products have received high ratings from thousands of satisfied customers, and their online forum offers a first-class insight into the neuroses of modern manhood: "Is there a risk that the pheromones will make me attract gay men too?" "Does it work even if you are not very handsome or muscular?" "Does the product come in a discreet package?" One attentive user noted that the pheromones do not work when meeting online. Another gave the product a low rating, noting with painful sincerity that he still didn't even seem to be able to attract a fly. Of course, the high average consumer ratings in no way prove that the pheromone method works, but may instead be a placebo effect—because customers *believe* the product has an effect, they behave more confidently and interpret the behavior of others differently than they would otherwise.

In today's appearance-obsessed culture, the pheromone concept offers some unique business opportunities. Tired of unsuccessful dates, artist Judith Prays invented the "Pheromone Party," a type of singles' get-together where people were allowed to get to know each other's smells in advance via used T-shirts sent through the mail. The idea was that the smells would reveal whether or not they were compatible. It attracted a lot of media attention and perhaps someone found their true love in this innovative way. However, the name Pheromone Party was possibly misleading as body odors are not the same as pheromones. Attraction to someone's body odor doesn't have to have anything to do with pheromones at all, but can be due to simply liking the smell for some reason. Judging by a website that hasn't been updated for a long time, the business seems to have ceased in 2014. The same goes for several other pheromone companies that have sold products based on the promise of pheromone attraction. They quickly emerge as a fun new product, often gaining media attention, but interest then quickly fades. It seems difficult to build a profitable business based on pheromones, despite the cultural popularity that has emerged over the past few decades. If the purported pheromones could really influence people's behavior and create attractiveness to the opposite sex, a significant industry would probably have emerged by now.

Are androstadienone and estratetraenol real human pheromones? For those who are persuaded by popular science articles, this may be self-evident. For example, many people are familiar with the research that women who live together synchronize their menstrual cycles. The implication is that they have an equal chance of being fertilized by a visitor with good genes. But it turns out that the conclusion that many have drawn about the existence of pheromones has in fact been premature. I met Oxford professor Tristram Wyatt at a conference in Tokyo, where he explained in a low-key but incisive way exactly what the problem is with human pheromone research.

It began in earnest in 1991, when the fragrance company EROX announced that it had patented androstadienone and estratetraenol as "possible human pheromones." The claim was curious, as no evidence was provided as to why these particular molecules should be described as pheromones. At a conference the same year in Paris, scientists presented a research study in which they used these molecules to cause activity in small biological samples from the so-called vomeronasal organ, sometimes called Jacobson's organ, located in the nasal cavity, which in many other animals acts as a second sense of smell specializing in pheromones. In humans, however, the vomeronasal organ is an evolutionary leftover—there are no nerve fibers connecting it to the brain! Research on human pheromones could have stopped at this anticlimax, but instead it took off again in 2000. A paper was published by renowned olfactory scientist Martha McClintock and her colleagues at the University of Chicago. The researchers showed that the molecules described as pheromones by EROX appeared to increase the positive emotional state of female, but not male, test participants. The effect was the same for both the "male" androstadienone and the "female" estratetraenol. The researchers were cautious in their interpretation of the results, considering that there were not yet sufficient reasons to call these molecules—donated to them by EROX—pheromones. Nevertheless, the article led to a flurry of research projects, newspaper articles and companies whose main idea was that human pheromones had now been discovered.

Around the same time, McClintock and her colleagues published a study in the journal *Nature,* claiming to have found "definitive evidence" for human pheromones. They argued that they had created a synchronized menstrual cycle in women by stimulating them with arm sweat from other women, collected at different stages of the menstrual cycle. Participants wore a small piece of cloth from the donors' armpit on their upper lip. Smelling the sweat collected at the follicular phase before ovulation shortened the participants' menstrual cycle, while sweat collected at ovulation itself lengthened their menstrual phase. Although McClintock and her colleagues could not say exactly which molecules in the sweat affected the menstrual cycle, it was considered a major advance.

Since then, over forty scientific studies have been published using one of the molecules patented by EROX. Most have found some kind of effect on mental states, behavior or physiological processes. On the surface, then, human pheromone research may appear to be a success story. But recent developments in the field of psychology put these research papers in a very different light.

Pheromone research faces many of the same problems that began to be discussed by psychologists after 2011. At that time, researcher Daryl Bem published a remarkable article in the prestigious *Journal of Personality and Social Psychology* titled "Feeling the Future." He presented results from nine experiments, all of which pointed to a startling conclusion. People were capable of predicting future events, such as that the next image to be randomly displayed on a monitor would have an erotic or violent content. The scientific community reacted strongly to Bem's article. How was it possible for an article to be published when the conclusion was obviously absurd and not based on a scientific theory? After all, predicting the future is a supernatural ability that goes against everything we know about the laws of the universe. The debate revealed that Bem's research had serious problems. He had conducted many experiments over the years, but most of them did not produce interesting results. The public was only allowed to read about Bem's most sensational results, which were probably due to chance but gave a misleading impression when presented together.

Around this time, similar doubt was also directed at brain-imaging research, which was criticized for using sloppy statistical methods. Some researchers showed that such sloppy methods could even show "brain activity" in a dead salmon. Others showed that the results presented as truths in top science journals were actually too good to be true. Psychological tests and brain activity measurements were simply not stable enough to produce such perfect results time and again, especially since the number of participants in the studies was often quite small. So something was fundamentally wrong with how the research results were produced, and with the journals that published them.

In recent years, the picture has become clearer. We now know that many, perhaps even most, psychological research results are not repeatable, or at least weaker than previously thought. Some researchers, out of ignorance or recklessness, have used statistical tricks that border on malpractice, which can increase the risk of a false result from 5 to 50 percent. Daryl Bem is one of the most notorious examples, but there are similar cases in many fields. Add to this a publishing culture that is driven by commercial interests and emphasizes novelty at all costs, and the increased competition in the scientific community that makes sensational research results a must in order to win reputation and research grants. Too few have been interested in replicating their own or others' results to see if they are really reliable. The good news is that many researchers, in psychology and other fields, have recently changed their methods to encourage replication.

Thus, we have learned to look with suspicion at research topics where only sensational results are newsworthy, where the number of participants is low, where the theoretical foundations are weak, where the most important discoveries have been criticized for inadequate methods and where exact repetitions are rarely made to confirm one's own or others' research results. In the research on human pheromones, all these warning bells are ringing, according to Tristram Wyatt and other critics. Researchers have presented several striking findings on how the alleged human pheromones affect mood, behavior and brain activity. But when you look closer, the skepticism increases, because there are not many results that are repeated in multiple independent studies. Each researcher simply seems to find different things. Even the famous studies on synchronizing menstrual cycles by McClintock have been criticized for statistical problems that may have influenced the sensational results, and researchers disagree on how to interpret them. The risk of publication bias is very high in this research field. It is difficult to estimate how many failed pheromone results are sitting in researchers' drawers, but it is likely to be many. Even if the pheromone research had been robust and repeatable, other problems would have been difficult to overlook. The purported pheromone molecules, androstadienone and estratetraenol, were not used by the researchers in the realistic concentrations found in human sweat, but in hundredfold-higher concentrations. Therefore, even the most exciting results become a dubious description of reality. In my view, there is still no strong evidence for human pheromones.

Instead, the research field is moving in another, more promising direction: collecting real body fluids and presenting their odors to participants in experimental situations. In such experiments, hundreds of odorants are mixed in unknown concentrations, and it is not known which molecule or molecules are involved, but it is at least a more realistic situation.

Some researchers have begun to focus on body fluids other than armpit sweat, because there is really no reason why pheromone molecules should only be found in the armpits, when there is just as much smell from breath, feet or crotch. Perhaps it is a certain convenience that has led researchers to focus so narrowly on armpit sweat. Nowadays they also study how we react when we smell tears, or disgusting smells such as urine and feces. These are not usually referred to as pheromones, but rather as *chemosignals*. The term chemosignals is more permissive, taking into account that, as with most human psychology, the possible effects the chemosignals might have are influenced by learning and are situationally dependent rather than automatic, as with the original insect pheromones.

Some of the hottest research on human chemosignals involves motherhood. Newborn babies have a well-developed sense of smell that helps them find their mother's breast. There are special glands around the nipple that secrete an odorant. The baby's sucking reflex can be triggered by this smell alone. Is this evidence that the substance is a pheromone? No, most likely the baby learns to associate the smell with the milk. According to the original definition of pheromones, the pheromone effect should not be created by learned associations. But even if it is learned, the smell of the mother is a strong signal for the baby. And the same goes for the baby's smell—it contains thirty-seven different odor molecules. Some of these molecules are only secreted by newborns, which is perhaps why we find their smell so special and wonderful. It creates a very strong reaction in the brain's reward system in new mothers. Pheromones or not, smells create strong emotional bonds between baby and parent. The sense of smell is a strong argument for both parents to have close physical contact with the newborn.

Research on human pheromones has focused unilaterally on sexuality, which is a limitation. Human urges are probably not affected by pheromones in the same way as some other animals are affected. But sex is an activity that involves most of the senses, and smell and taste obviously play important roles when lovers are close to each other. Research shows that the sense of smell is particularly important for our sex lives and love relationships. Researcher Anna Blomkvist focused her PhD work on how the sense of smell affects our love relationships and how we create emotional and social bonds. In one study, she gave her research participants small electric shocks during the experiment, which led to a stress reaction that could be measured by increased sweating on the skin. But the participants were also allowed to smell their partner's body odor as a way to reduce the stress response. It worked; women had a lower stress response when they smelled their partner's sweat, at least compared to when they smelled their own sweat. But the effect was not equally strong for everyone. The stress-reducing effect was strongest in women who had a secure relationship with their partner. Thus, our partner's smell can provide security in stressful situations—if we associate the smell with security.

In a follow-up study, Blomkvist designed a large survey where over a thousand participants answered questions about their sense of smell and their private lives. Her results support the idea that the sense of smell is important for one's love life. Those who reported having a sensitive sense of smell were more satisfied with their sex life and less likely to cheat on their partners. Does this mean that smell training could lead to a better sex life? Yes, it is possible. But maybe it's really more about being *aware* of the smells around us and letting them affect us. We know that people who say they have a good sense of smell don't always have as sensitive noses as they think, but they react more strongly than others to the smells they smell. So the sensitivity is emotional and mental, rather than purely sensory. This may explain why they have a better love life—they have a kind of sensuality based on an awareness of smells and their emotional significance.

In contrast, people with a poor sense of smell sometimes have difficulty forming close relationships. This is a problem that is rarely acknowledged. When we go on a date, the meeting often revolves around food and drink. This is probably not a coincidence. Because the sense of smell is so personal, every discussion about food and drink is also a way to quickly connect with each other. Which dishes on the menu look the best? Which wine do you choose? Maybe a beer? Or water? Talking about food, smells and tastes is a bit like talking about the weather. It's an easy icebreaker with people you don't know very well. But at the same time, our food preferences signal something about who we are. Person A, who enjoys spicy food and unexpected, experimental flavor combinations, may have a different personality and values than Person B, who prefers mildly spiced food and familiar dishes. Guess which of them is more likely to have traveled around the world, gone skydiving, had many acquaintances and more sexual experiences—or had a concussion? These seemingly disparate behaviors and experiences are common among sensation-seeking individuals, who also often enjoy spicy and unusual or experimental foods. Who we are is reflected in what we eat, in other words. But for someone who has no functioning sense of smell at all, what's on the plate matters less. It becomes difficult to talk about food and drink. And it becomes quite challenging to cook for someone else. The social community that arises at mealtimes has been partially lost.

Research from around the world points to the personal problems that arise when the sense of smell is impaired or lost. Researchers from Iran reported on patients with loss of smell caused by head injury. Among those with olfactory impairment, depressive symptoms were more common, and the greater the damage to the sense of smell, the more common the depression. While only 6 percent of the control group had severe depression, 25 percent of the olfactory patients had severe depression. Similar results emerged when they examined whether the participants were sexually satisfied in their love relationship. Among the control group, 67 percent were completely satisfied with their sex life and only 12 percent were dissatisfied. Among the odor patients, only 22 percent were completely satisfied and 32 percent were dissatisfied. When it came to how satisfied the participants were with their relationship in general, the difference was similar: 8 percent of the control group but 21 percent of the odor patients were dissatisfied with the relationship.

Impaired sense of smell is detrimental to one's sex life, but researchers do not agree on exactly why. One theory is that loss of smell affects sexual and love relationships because it leads to depression; the depression is what creates the relationship problems. But it could also be partly the other way around: that depressed people have a poorer ability to smell because the brain is in a kind of imbalance—it is well known that memory abilities are often affected by depression. According to this interpretation, it is the sense of smell that is at the heart of the problem, affecting both our ability to enjoy sex and to be happy in general. Because sexual problems are considered an embarrassing thing to talk about, patients sometimes avoid telling their doctors. But for those physicians who encounter patients with olfactory impairment, it is important to know that it can erode quality of life in many differ-ent ways. And those who suffer from olfactory impairment may not be aware that it can be the cause of their love life faltering—but this insight can help couples affected by olfactory loss to improve their relationship.

Body odors can bring pleasure or well-being, but just as often they are repulsive. Unlike pheromones, disgusting body odors are almost taboo. We rarely talk about them, and until quite recently we knew very little about their psychological effects. In recent years, research has shown that our body odors, but also our feelings of dis-gust toward them, can reveal things about us—things that we would prefer to hide.

CHAPTER 8

DIRTY SECRETS

THE WOMAN AT the ice cream counter looks at me skeptically. "Are you *really* sure you want it?" I nod. "Usually only us Asians enjoy durian ice cream." I say, somewhat bombastically, that I'm willing to take the risk. I am in Singapore to attend the Organization for Human Brain Mapping conference. One of my best experiences from the trip to Singapore is smell-related: the amazing Flower Dome, a giant flower garden enclosed in a glass dome where amazing smells and colors wash over the visitor. But my worst Singaporean experience also has to do with the sense of smell. Shortly after being seduced by flowers and other plants from all over the world, I stand outside the dome, about to make a big mistake. At the ice cream stand outside the flower garden, I order a cup of ice cream flavored with durian, the stinking fruit, or as it is called in Southeast Asia, the king of fruits. I thank the lady at the counter, grab a plastic spoon and put a big scoop in my mouth. The taste is indescribable. Some say it's like eating vanilla cream in a sewer. The celebrity chef Anthony Bourdain said that after eating durian fruit, your breath smells like you'd been tongue-kissing your dead grandmother. To me, it feels like the ice cream has been flavored with sweet, floral perfume mixed with a generous dose of fermented herring.

The smell of durian is a good example of how culturally influenced the sense of smell is. The durian fruit is considered a delicacy in parts of Southeast Asia. Yet it is banned on the metro and buses, perhaps partly out of concern for tourists. Is it natural to like the durian fruit? When *New York Times* Southeast Asia correspondent Thomas Fuller had Thai elephants, known for their keen sense of smell, try the durian fruit, their response was unequivocal. They crushed the melon-sized durian fruit and then devoured every last bit. My own durian experience, however, ended in defeat. I continued to eat my durian ice cream as I strolled around Singapore's beautiful parks with my colleague. But after a while, a growing nausea set in. Defeated by the durian, I threw about half the ice cream into a garbage can. Hours later, I could still smell the stench of my own breath and remembered Bourdain's words.

The allure of pheromones has put the spotlight on body odors that arouse desire and stimulate our sex drive. Yet, it sometimes seems that the most common reaction to body odors is forgotten—people stink. Unpleasant body odor comes with a lot of stigma. I recently heard an acquaintance of mine say that she absolutely did not want to recruit a colleague to the workplace because "he *stinks*!" Unfortunately, discovering if you smell bad is not as easy as you might think. We become numb, losing sensitivity to the smells we are constantly exposed to. Only when we smell new odors do we have full sensitivity. The best solution in most cases would probably be to tell someone when they smell bad, but this is not really socially acceptable and many people avoid the touchy subject. Instead, the person may walk around without understanding why he or she is avoided in the workplace. However, the problem is much worse for those who have lost or have a weakened sense of smell. Research shows that one-third of people who have lost their sense of smell walk around afraid of smelling bad themselves. The COVID-19 pandemic affected the sense of smell for up to half a million Swedes, and many of them were naturally worried about that. When everyone returned to in-person work after a long period of working from home, it's easy to imagine the stress of not being able to smell yourself.

No matter where we live in the world, bad smells evoke strong feelings of discomfort. The link between smells and disgust is so strong that some researchers believe it is the very point of having a sense of smell. Disgust is the brain's way of telling us that we risk getting sick if we put that stuff in our mouths. The reason why we react so strongly to the smell of sweat is also related to diseases. The smell of sweat can signal that a person is sick. The odor is often a result of the chemical by-products of the disease process that are "vented" through our skin. Why do diseases smell, and what is the purpose of being able to smell them? My colleague Mats Olsson, professor at Karolinska Institutet, argues that human psychology is designed in a way that allows us to avoid disease, rather than letting the immune system take the brunt of it when we do get infected. According to this perspective, we may have developed strategies to use our senses to perceive diseases, the situations or people that carry the risk of being infected, and behaviors to avoid them.

Olsson's research team uses *lipopolysaccharides* (LPS), substances that make participants temporarily ill. A few hours after an injection of LPS, their bodies secrete various substances that show that inflammation is taking place. These people's body odors are picked up by the T-shirts they wear, and these are later used in olfactory studies. Their body odor, the research shows, becomes strongly malodorous because of the temporary illness. However, it has been difficult to isolate the exact odorants secreted in the sweat. The illness also manifests itself in the face, in "droopy" eyes and pale skin, in their voice and in their tired gait. Participants who see these sick people's faces and smell their T-shirts are negatively affected and say they want to avoid them, even though they do not know they are sick, and in many cases would not be able to tell who is sick and who is healthy. The smell of illness, and something about the appearance of the sick, has a repulsive effect on us.

The emotional charge of smells, in its most primitive form, is about positive and negative smells. Smells we like make us a little happier, and smells we dislike make us disgusted. It doesn't take much imagination to think that an unpleasant person also has a disgusting body odor. Remember medieval devils signaling their presence with smelly farts, and the "smell of holiness" that surrounded saints even after death. Going a little deeper into the emotional life of the sense of smell, modern olfactory psychology research has shown that smells can also often be described as relaxing or activating, and these categories of smell are found in many different cultures. Think of lavender, which is often described as a stress-reducing aromatherapy smell, and the citrus tones often found in hygiene products that want to give a sporty feel. Other emotions evoked by smells are sensuality/desire and hunger/thirst. These correspond to two of our main survival drives—for sex and food. Thus, when we smell, we classify the smells, often without thinking about it, according to an emotional register that has only a few strings, but that has become deeply rooted in our brains over millions of years.

In many animal species, body odors provide important signals about the animal's genetic makeup, and therefore body odor—if interpreted correctly by other individuals—can reduce the risk of inbreeding. Does this also apply to humans? Researchers have argued that women can smell whether a man has a gene set that would benefit a child's immune system. But no, this sensational result, despite several attempts, has not been repeated and is therefore probably a myth.

The smell of sweat is unpleasant, and the less a person smells of sweat, the better. In this way, strong body odors act as a primitive signal to the recipient to stay away. The opposite is true for pleasant body odors, which make us attribute positive qualities to a person, and even judge them as slightly younger than we would otherwise. But the importance of smell in real life often depends on the situation. How well do we know the person, and what feelings do we already have for them? For a person we have already formed an opinion about, a less pleasant body odor may not make much of a difference, while on a first date it may be more important. The role of body odors also depends on who we are and how much importance we attach to smells and odors. Women tend to value good body odors more than men do. A study of American college students found that young women consider body odor to be more important than appearance when choosing a partner. For young men, on the other hand, looks were more important. These different priorities make sense from an evolutionary perspective. Smells can reveal sources of disease that can harm a fetus and lead to miscarriage. It is easy to imagine that women focus on their partner's smell because they want to avoid diseases and other "hidden defects." But the difference also, I suspect, tells a story about our cultural history. The sense of smell, as you now know, has, since ancient times, been considered a primitive sense, emotionally charged but useless as a tool for thinking. Women, sometimes considered less civilized than men, have been responsible for domestic childcare and cooking—the intimate, smelly world of the home. The idea that women have a sensitive sense of smell is part of our cultural heritage. The question of whether it is gender roles or biology that cause women and men to attach different importance to the sense of smell remains unanswered.

There are many reasons why some people have a strong smell, and you can influence your own smell. Our body odors are affected by what we eat. The main food that leads to a strong and unpleasant sweat smell is large amounts of red meat. Scientists have known for a long time—and this is probably obvious to most people—that men tend to smell more than women, and this is partly due to the fact that they eat more red meat. Therefore, it is actually quite easy to determine whether the person is a man or woman, or whether they are a carnivore or vegetarian, just by smelling their arm sweat. Other ingredients that affect our sweat smell are strong spices and garlic. Researcher Jitka Fialová has shown that eating garlic has effects on body odor—but not in the way you might think. Participants in Fialová's study ate bread with a few cloves of crushed garlic, and their body odor was collected on T-shirts and later presented to other participants. The smellers did not actually recognize the garlic in the sweat smell, but rated the odor as more pleasant and attractive than a "garlic-free sweat." The researchers believe this is due to the health-promoting effects of garlic, which can make us like the smell. Personally, I suspect it's more about a changing cultural convention. Previously, many people thought that you should avoid eating garlic if you were going to a party or other social events. "Smelling like garlic" was considered embarrassing and disgusting. But times are changing, garlic has become increasingly common in the food culture of Northern Europe, and perhaps we have simply learned to like the way it affects our body odor.

Those who smell bad are often unaware of it, as we are numb to our own smells. But some people suffer from the opposite problem— they think they constantly smell bad even though they don't. This is a rare psychiatric syndrome called *olfactory reference syndrome* (ORS). Sufferers are obsessed with their imagined odor and try to remove it in all possible ways, such as excessive hygiene practices, frequent use of chewing gum or mouthwash and unusually frequent changing and washing of clothes. People with ORS have a distorted perception of their own body odor much like eating disorders cause a distorted body image. The imagined smell is linked to anxiety and low self-esteem, and the various hygiene behaviors associated with ORS are related to obsessive-compulsive disorder.

In recent years, my research team has been exploring the psychology of disgusting body odors. This research has, perhaps unexpectedly, led us to big, sensitive political issues. Bad body odor is highly stigmatized in our culture, and a person who is considered to smell bad, perhaps due to homelessness and difficulty in maintaining hygiene or washing clothes, is likely to be treated badly, perhaps even shunned. In many cultures, bad body odor is considered reason enough to avoid a person. Yet it has been unknown why some people are so disgusted by smells that others can easily tolerate.

We already knew that people are very different in how they react to smells, as previous chapters have shown. And in our research on malodor, we chose to shift the focus—from those who emit odors to those who react to them. The first question we asked ourselves was whether people who were particularly sensitive to body odor also had political attitudes that were different from the average. We soon realized that it would be very impractical to invite hundreds of people to the research lab to smell sweat, pee and poop and other body odors that are expected to induce disgust. We were inspired by other disgust researchers—yes, there are quite a few—who have developed questionnaires that ask participants to rate how disgusted they are by various events (such as seeing someone eating vanilla ice cream with ketchup on it, or licking a used but cleaned fly swatter!). We described different situations related to disgusting body-related odors such as sweat and poop. An example of such a situation was: "You use a public toilet after a stranger. It smells strongly of their poop." The participant was then asked to rate how disgusting it would be on a scale of 1 to 5. Not surprisingly, many people gave high ratings. To confirm that the questionnaire was indeed indicative of how people react to body odors, we first compared the results of the questionnaire with disgusting reactions to real sweat odors. Once we had confirmed that the questionnaire gave reliable results, we could use it without having to expose the participants to sweat odors.

Over the past years, thousands of participants have anonymously answered the questions in our online questionnaires. The results are clear: there is a firm link between odor disgust and political attitudes. Body odor disgust is strongest among people who want a society that is governed by a strong leader, where old traditions and gender roles remain unchanged and where people who break the law and societal norms are severely punished. It is a society that does not change quickly, and where different groups have different roles. Researchers have a term for such attitudes: authoritarianism. What does this have to do with disgusting smells? Our research gradually gave us an increasingly deepened insight into the psychology of disgusting smells.

Feelings of disgust about body odors turn out to reveal our deep-seated fears, values and views of other people. People who are easily disgusted—let's call them "thin-skinned" for now—tend to dislike contact with strangers, and they are particularly afraid of contracting diseases. When judging other people's faces, they often think that people look sick, even if they are perfectly healthy. They also often think that different groups of people should be kept apart and not mix with each other. The latter is a convenient solution for those who do not want to be exposed to viruses, as viruses do the most damage when they spread to new groups that do not already have immunity. We initially focused on US trial participants for the 2016 presidential election. We found that the thin-skinned tended to like Donald Trump rather than his main opponent, Hillary Clinton. They also have a slightly different view of morality compared to those who tolerate strong body odors. They feel more disgusted when respected symbols are desecrated, such as when graffiti is spray-painted on a church. Marta Zakrzewska did her PhD in my research group, focusing particularly on the psychology of disgust. She asked participants what they thought about refugees coming from a foreign country—a hypothetical Central African country that we call Drashnee. Refugee reception is a sensitive issue and the answers say a lot about our feelings and values, even if in the case of this particular study, we use an imaginary group. Some participants are negative about this imaginary group of refugees, while others are positive and want them to enter the country. The participants who are most disgusted by smells are also the ones who have the most negative attitude toward the refugees. Initially, we found the correlation in American participants. Then we repeated the experiments with Swedish and Italian participants, with exactly the same results in both countries. In a new study, we looked at whether the relationship also applied to study participants living in Asia, South America, Australia and Africa. In all continents we found similar results. The participants who were most disgusted by body odor were also more negative toward refugees. To investigate whether there was something special about attitudes toward African refugees in particular, we added a hypothetical refugee group from Eastern Europe. But it turned out that even for this group, which most participants considered more culturally similar to themselves, the familiar relationship between xenophobia and disgust sensitivity was found. The research has thus led to the conclusion that people with strong disgust sensitivity to body odors have a kind of fear of infection. This leads them to want to create a society that minimizes the risk of contagion, even if it leads to reduced mobility, dynamism, openness and innovation. They are more likely to distance themselves from perceived "others," such as people they have never met.

Since our sense of smell often causes rejection, it is not surprising that racism often manifests itself in descriptions of body odors. During the age of slavery, Europeans were committed to the theory of miasma, that smells could spread disease. In European cities, smells were gradually declining, thanks to improved sewage systems and access to running water in homes. It was therefore timely to portray Africa and its people as smelly. Historian Andrew Kettler argues that during the seventeenth century, it became increasingly common for Englishmen to read descriptions of how the entire African continent smelled as if it were poisoned, and that its inhabitants smelled terribly. Colonists and travelers who followed the Atlantic coast of Africa spread the image of a dangerous, diseased and undeveloped continent. It gave readers the impression that Africans were not only different, but also repulsive and less human. This was not a new idea; similar descriptions had been used by Christians to justify the persecution of Jews. These descriptions became more common as the transatlantic slave trade expanded. Kettler calls it *olfactory racism*, an expression of the slavers' own prejudices and interests. It helped to spread the misconception that Africans and Europeans belonged to different biological races, which became increasingly popular in the eighteenth century and was used to defend the slave trade. Since the sense of smell is shaped by our preconceptions, it could be used to confirm preexisting prejudices. The sense of smell thus became a powerful emotional tool in the service of slavery.

The link between smell and disgust is strong. Just think about that dish you were forced to eat as a child and how the smell still makes you feel uncomfortable. Or those mussels that seemed so good but made you sick, and you can't bear to eat them ever since. For humans, disgust has led us to avoid the unknown. We almost always think that unfamiliar smells are bad. But the other side of the coin is that we quickly get used to them—we can even learn to like the smell of sour herring. Our nose remembers that last year's surströmming party went well. Disgust for smells is thus linked to the fear of the unknown and what has made us sick in the past. This ancient biological function has been preserved over millions of years. Even rats react negatively to the smell of food that has made them sick and refuse to eat it again.

Is it immoral to dislike disgusting smells, and should we just adapt to any smell in the environment? No, but we should try to be more aware of our reactions. Our strong feelings of disgust can be used for manipulative purposes. In 2010, conservative New York politician Carl Paladino sent out a flyer impregnated with the smell of rotting garbage. "Get rid of the stench of corruption—throw away the garbage" was Paladino's message, and the garbage he was referring to was his liberal opponents. In a way, perhaps, he was ahead of his time. On social media, it is now not uncommon for political opponents to be called disgusting or repulsive. Political hate speech often involves associating opponents and vulnerable groups with pests and other disgusting things. It is a way of degrading the humanity of the group and, in some cases, justifying their persecution. Carl Paladino's sophomoric flyers did not take him all the way to the governor's office, but a few years later Donald Trump was elected president of the United States, after describing Mexican immigrants as spreading disease, and hate crimes against Latinos and other minorities increased. The sense of disgust aroused by bad smells has helped keep people alive throughout history. But in modern, open societies, this fear of contagion comes at a high price—thin-skinned people risk living more restricted lives, and their strong emotions can have destructive results. By studying the sense of smell, we believe we can learn more about how these strong emotions work, and how disgusted reactions to smells are linked to other personality traits. Research suggests that both prejudice and authoritarianism may be linked to the brain's olfactory system. This may explain why these views are often based on strong emotions rather than facts. But when we raise our awareness of smells, and the emotions they evoke in us, we can build up a resistance to those who want to exploit those emotions for destructive purposes.

The idea that feelings of disgust are ancient, biological warning signals to keep us alive may seem to contradict the main idea of this book, that the sense of smell is an interpretative, intelligent sense in which we use all our human brainpower. But even our strong feelings of disgust are largely learned and a result of interpretation and association. This is revealed when we examine differences between cultures. For example, anthropologist Constance Classen has described how among the Dassanech people of Ethiopia, the smell of manure is considered to make a man attractive. Why would they want to spread such smells around them? It is actually quite easy to understand. They are completely dependent on their livestock, and the smell of the animals signals wealth, status and the ability to provide for many children. For a Dassanech woman, the smell of manure means that the man is probably a good catch. Some may find the olfactory tradition of the Dassanech people strange. But no one is born disliking the smell of manure. A small child is not born with any strong aversion to particular smells at all. Instead, children learn from their environment to prefer certain smells over others. In our own culture, we have the habit of rubbing whale vomit on our skin. Ambergris, a substance found in whale intestines and ejected through the mouth, is used in our expensive perfumes. Anal secretions are also desirable; the perfume industry has long been interested in the anal glands of the musk deer, the civet cat and the beaver. So we are hardly strangers to smells with a filthy origin. Which smells we love and hate depend largely on the associations, thoughts and feelings they evoke.

The sense of smell is a powerful biological tool to reduce our risk of disease. Feelings of disgust distance us from the potential risk of infection. Food smells that evoke bodily memories of past nausea, but also unfamiliar smells, or too strong smells, can disgust us. The disgust aroused by smells is an ancient alarm system that is sensitive but often activated unnecessarily. We now know that the miasma theory was wrong; it was not the smells that made us sick, but bacteria and viruses. This is why the cognitive perspective is important. It can teach us to reflect more on our emotional reactions to smells and try to reduce the stigmatization of abnormal body odors. They are not only a reflection of our cleanliness or lack thereof but can sometimes be a result of chronic diseases or hormonal changes—such smells are difficult to wash away. I hope that an increased knowledge of the sense of smell will lead to an understanding of the smells that we instinctively find disgusting. Our feelings of disgust at a strong smell are not predetermined and may even be reinforced by our prejudices toward the unknown and the different. We can learn to tolerate smells that we find unpleasant, and even if we don't learn to love the smell of other people's sweat, durian fruit or fermented herring, it's a good idea to at least try to be more accepting of them. In the remaining chapters, I will describe how we can rediscover our sense of smell, and how it can change our lives.

PART III

LEARNING TO SMELL AGAIN

CHAPTER 9

SMELLS AND LIFE

I MAGINE A CHESS game where all the pieces look the same. There are differences between the pieces—but you can only understand them through the pieces' unique smells. In this game, the pawns smell like sage, the bishops like basil, the king like cinnamon. Such a game was developed by artist Takako Saito in the 1960s. She created various "distorted chess games" to challenge our notion that art is something we passively observe with our eyes. Playing a smelly chess game is an interactive piece of art where you have to use multiple senses and concentrate deeply. Would you be able to play such a game? The concept of a smelly chess game is fascinating and raises many questions for the olfactory connoisseur. What are the limits of the problems we can solve with the sense of smell? And can the sense of smell help us keep our brains vital in old age?

As you now know, we humans usually have an excellent and sensitive sense of smell. Unfortunately, this is not the case for everyone. The fragile structure of the sense of smell—the receptors are exposed in the mucous membrane—makes it vulnerable to viruses and bacteria that can quickly kill or slowly wear down the sense of smell. The wear and tear increases over time, as our colds, concussions and life's other harmful events are added to the effects that come with age in the form of a drier mucous membrane and fewer branches on the nerve fibers. A weakened sense of smell therefore most often affects the elderly. About half of all elderly people have a weak sense of smell, while only one in six middle-aged people suffer from it. Loss of smell is a major but hidden disability. And unlike a hearing or visual impairment, there are no aids for the sense of smell, no equivalent of hearing aids or reading glasses to help the nose correct for the impairment. In the elderly, it is often suspected that the weakening of the sense of smell has been creeping up for many years, so when researchers ask older people about it, the vast majority say their ability to smell is normal, even though an examination shows a clear impairment in half of them.

When the sense of smell is weakened, it is rarely a total elimination, but a gray scale, where most of the smells remain, but in a reduced form. Nuances and faint smells are lost and the weakening leads to problems. Older people often find that food tastes less than it did before. This is a problem in itself—good food is one of the most important pleasures of old age. But not being able to sense the aromas of food can also lead to a reduced desire to eat. Because older people are often unaware that they have a weakened sense of smell, they may not understand what is causing their lack of appetite. Having a poor appetite is of course bad, but it is not always a health risk. In fact, many older people continue to eat normal amounts of food out of habit, despite a loss of appetite. But for some people, loss of smell puts them at risk of malnutrition. Those who have a loss of smell have less variety in their diet and often eat the same food several times a week. This is because the brain engages the sense of smell to determine what kind of food, and how much, we need to eat. The smells of food set off a cascade of biological processes in the body. They create a release of insulin in the blood, which lowers blood sugar levels and makes us hungry. Smells also increase the production of saliva and stomach acids, and they increase body temperature and heart rate. In other words, the body prepares to break down and absorb nutrients. Without a good sense of smell, the whole system is thrown into disarray.

The tempting smell of food stimulates the brain's reward system, but this is quickly blunted once we start eating—we feel full from that particular dish and move on to something else, such as a dessert or a snack. In elderly people with a weakened sense of smell, both hunger and this "sense-specific satiety" are reduced, so they don't enjoy the food and its variety as much, nor are they as interested in taking more salad, refilling their drink or having a dessert during the meal. There is therefore a risk that the weakened sense of smell leads to poorer nutritional intake and reduced body weight. This is a serious problem. Malnutrition is one of the most dangerous things that can happen to older people. Those who already have other health problems, and who live in residential and nursing homes, are particularly affected. Perhaps the olfactory engineers of the future will become better at creating food and drinks that taste and smell good for older people with olfactory impairments. Research has already shown that it is possible to compensate for loss of smell. The most common recommendation is to increase seasoning and enhance flavors. One way is to use glutamate, the flavor enhancer found in some spice mixes and herbal salts. Elderly people who received such enhanced food in controlled research studies experienced healthy weight gain. Unfortunately, there are many prejudices about glutamate, and some people believe that it is unhealthy or even toxic. This is not true. Glutamate occurs naturally in many common foods such as tomatoes and cheeses and it is not dangerous. It is true that a small proportion of the population is particularly sensitive to glutamate—they get nausea and headaches—and they should avoid such flavor enhancers. But for the vast majority, glutamate can be a good addition to make food more palatable.

Appetite can be stimulated by engaging all the senses. According to the cognitive perspective, our sense of smell is influenced by expectations, so the color, taste and texture of food can actually compensate for a weakened sense of smell. Making food colorful, tasty and crunchy can thus be a recipe for fighting malnutrition. Researcher Alexander Fjældstad and his colleagues at Aarhus University have started cooking classes for people with loss of smell. Many of them lost their sense of smell during the coronavirus pandemic. The courses teach participants what works for them and what does not. An important trick is to make use of the inherent tendency of the sense of smell to blend sensory impressions from different parts of the brain and create an overall impression. Fjældstad has suggested that people with loss of smell compensate by enhancing other sensations. These are summarized in the "four T's": *temperature* (mixing hot and cold dishes), *texture* (such as crispness), *tactility* (such as eating with your hands) and *trigeminality* (stinging, cooling or hot spices that stimulate the trigeminal nerve). Many patients bring family members to the cooking classes at Aarhus. It is difficult to understand what it is like to live without smell unless you have experienced it yourself, and the cooking classes have been a success.

In recent years, the sense of smell has taken on a different role among psychology and brain researchers, as a window into the aging brain. One person who noticed that her sense of smell gradually deteriorated with age was Ingrid Popa, who lived in the Romanian village of Mühlbach. She could no longer experience the spiciness of food or the smell of flowers. She talked about it with her family and doctors, and among her friends, but was told it was natural, something that happened in old age. After a while, however, her memory and orientation began to deteriorate. A few years after losing her sense of smell, she was eventually diagnosed with Alzheimer's dementia. She is an example of how the sense of smell can testify to the health of the aging brain and help us anticipate—and perhaps prevent—its diseases.

My interest in the sense of smell was sparked when, as a student, I started working with Professor Steven Nordin at Umeå University, who specialized in the sense of smell and how it related to Alzheimer's disease. Together with Claire Murphy, one of the pioneers of olfactory psychology and a professor at San Diego State University, Steven had found that olfactory impairment was evident in people with a mild degree of dementia, but also in people who had a gene variant that quadrupled the risk of Alzheimer's disease. At the time, I was working as a nursing assistant in a nursing home where I saw the devastating effects of dementia. We asked ourselves whether the sense of smell could be measured to understand which elderly people would later develop dementia. If so, olfactory testing could become an important element in dementia assessments, guiding medical treatments that alleviate symptoms and thus helping those who would otherwise be affected. It became a topic that has kept me busy since I started my postgraduate studies. A weakness of the research at that time was that it was often based on small groups of patients already diagnosed with dementia. Thus, it was unclear whether an impaired sense of smell could actually be used to predict future dementia in the general population. I used data from the Betula study, a large-scale project started in 1988 in Umeå under the leadership of Lars-Göran Nilsson, which recruited thousands of participants from the population register to ensure that the group would reflect the population. A similar study, the Swedish National study on Aging and Care in Kungsholmen (SNAC-K), began around the same time in Stockholm. Fortunately, both these studies included tests of the sense of smell. Thanks to the Betula study and SNAC-K, my colleagues and I have been able to prove that the sense of smell can indeed be used to predict dementia.

My methods for understanding the sense of smell are based on experimental psychological research. These methods do not always have a direct medical benefit, but may eventually lead to the use of smell tests in medical contexts. The research maps, for example, how different tests of the sense of smell work and which parts of the brain the tests engage. Sometimes we need to invent completely new ways to examine the sense of smell. It does not have just one function. Like the sense of sight, the sense of smell can do many things: detect weak smells (detection), distinguish between different smells (discrimination), remember them (recognition) or match them to the right name (identification). Identification in particular has become a common way of testing the sense of smell. The person is asked to smell everyday odors and try, using different response options, to say what it is they are smelling. The smells are presented by the doctor or researcher, for example using special olfactory pens that have been tested for clinical use. On the basis of the number of correct results, it is possible to determine whether or not the person has an olfactory impairment. Unfortunately, these methods are often absent in Swedish clinics, even though they are easy to obtain and simple to use. I am often contacted by people who tell me that not even the ear, nose and throat clinic at their local hospital has proper smell tests!

An important, and surprising, insight from olfactory measurements is that they usually do not follow the same pattern as common cognitive and neuropsychological tests that measure long-term memory, short-term memory, spatial ability, etc. While the common neuropsychological tests are often interrelated, so that those who perform well on one of the tests usually do well on the others, the sense of smell seems to be largely governed by its own laws. A smart and vital person does not have to be very good at identifying smells. We often meet participants who have an excellent memory, but are shocked by how difficult it is to name ordinary, everyday smells. In scientific terms, this is called *dissociation*. Thus, the sense of smell can provide unique insights into brain functions.

The gene variant that leads to an increased risk of Alzheimer's disease is called APOE-ε4. My research showed that older people who carry this variant (25 percent of the US population) also have a reduced ability to identify smells. But even though these people were born with the gene variant, we saw that it only affected their sense of smell once they passed retirement age. This was a clue. The APOE gene is particularly important for the aging brain. It produces a protein that helps shuttle cholesterol between brain cells, building and strengthening them in various ways. Researchers have discovered several roles for the APOE gene, including influencing brain blood flow and nerve cell connections and signaling. One particularly important role for APOE is to remove beta-amyloid, a type of biological debris that gradually builds up in the brain as we age and risks destroying the brain's neurons. The different versions of the APOE gene, ε2, ε3, and ε4, are differently good at these tasks, with ε4 being the worst in class. Therefore, people with APOE-ε4 are four times more likely to develop Alzheimer's disease than the majority of people with ε3, and ten times more likely than those with ε2. If you have several people in your family who have developed dementia, you might belong to the part of the population that carries ε4. This does not necessarily mean that you will suffer from dementia. Even in the small group who, due to high Alzheimer's risk on both the mother's and father's side, carry double variants of the ε4 gene, 75 percent of them manage to avoid getting Alzheimer's disease.

Our results strengthened our hypothesis that olfactory impairment might be related to early disease processes in the brain. In my doctoral thesis, I showed that people who had olfactory impairment in combination with APOE-ε4 had a faster age-related decline in their cognitive ability. In the five years following the smell test, they had lost twice as many points in a test of general cognitive ability as the same-aged participants with only one of these risk factors.

Over the past decade, research on the link between loss of smell and dementia has accelerated. We Swedish researchers have benefited greatly from the Betula project and SNAC-K, and my research group at Stockholm University is focusing on how the human sense of smell can help us understand brain aging. Because the research projects span decades, we can follow the development of the participants as they age. In her doctoral thesis, my PhD student Ingrid Ekström showed that people who have an olfactory impairment not only have a higher risk of developing a faster decline in cognitive abilities, but also have an increased risk of actually receiving a dementia diagnosis within the next decade. Her interest in the subject has a personal background: she is named after her grandmother, Ingrid Popa, whom you met earlier in the chapter. Ingrid Ekström found that older people with intact cognitive abilities can sometimes notice a marked deterioration in their sense of smell, and the perceived deterioration is an indication that the risk of dementia is particularly high. Just like with memory loss, they could go to the doctor and find out if there is a risk of dementia. But unfortunately, there is not much a doctor can do in the early stages. Anti-dementia drugs have made great strides, but we are still far from being able to safely slow the progression of the disease over time. Trying to live healthily, keeping your brain and body in good shape and not worrying too much about tomorrow is often good advice in everyday life.

Sometimes people with a poor sense of smell contact me and say they are afraid of getting dementia. But a poor sense of smell is not necessarily a cause for concern. While it is true that the sense of smell deteriorates in the early stages of dementia, it is affected by many other factors than just dementia. Some deterioration is normal in old age and could be due to other causes. Instead, I usually describe it like this: those who retain a sensitive sense of smell into retirement can feel a little more relaxed, because dementia is rare among them.

Different brain diseases can affect the sense of smell in different ways. Alzheimer's patients often retain the ability to detect smells but lose the ability to identify them, causing them, for example, to guess "lemon" even though they smell cinnamon. Parkinson's disease has a very different effect. These patients can hardly smell anything at all. The loss of smell in Parkinson's affects 90 percent of patients, compared to the well-known tremor, which actually only affects about 75 percent. A recent Parkinson's patient told me that she lost her sense of smell twenty years ago, long before the other symptoms appeared. Smell tests may become increasingly important for understanding the onset of Parkinson's disease—and slowing it down at an early stage.

In recent years, we have learned much more about how the sense of smell is affected by Alzheimer's disease. We now know that the disease often starts in the regions of the medial temporal lobe and lower frontal lobe, both of which process smells. These regions also control memory storage, so the loss of smell combined with memory loss is particularly ominous. People who have APOE-ε4 are, as mentioned, particularly vulnerable. In these individuals, the signs of the disease—neurofibrillary tangles and amyloid plaques—are particularly concentrated in these regions. Together with neuroscientist Donald Wilson and his team at New York University, I was able to observe a direct link between the ε4 variant and the sense of smell. Wilson's team used mice that had been bred with the human APOE-ε4 gene but were otherwise identical to other mice, both genetically and in their environment. The ε4 mice reacted completely differently to odors. We put a small lemon-scented cotton swab in the cage of each mouse, and all the mice came out and sniffed curiously at this exciting smell for about five seconds. After a while, we tried again. The mice started sniffing, but only for a few seconds, then they lost interest, as it was the same smell as before. All except the ε4 mice—they kept sniffing as if they had never smelled it before. In later experiments, we could see that the ε4 mice had chaotic, excessive activity in the olfactory brain neurons. This was probably the reason why the mice could not remember the smells they had previously sniffed, as chaotic activity in the olfactory brain makes it difficult to distinguish between new and old smells. Thanks to these mice, and because their olfactory brains are quite similar to those of humans, we might understand better why people with APOE-ε4 have a poorer sense of smell.

Brain training has long fascinated those in the field of psychology and brain researchers. Researchers have increasingly abandoned old preconceptions about the limitations of the brain. Brain capacity is not entirely fixed and innate, but can be developed throughout life. Research shows that education leads to a gradual increase in intelligence. And we are never too old to learn. Learning a new language, doing crossword puzzles, playing the piano—those who stimulate their brains have a better chance of living a long and healthy life. This is especially true if you also exercise and live a healthy lifestyle. Despite the fact that the world's population is getting older, this has not resulted in the explosion of dementia and health care costs that many had predicted. This is because in many countries, older people today have much better brain functions than those of fifty years ago. They have better food, better education and health care, and healthier jobs and lifestyles. Our genes cannot have changed much in this short time, so the change must be caused by changes in our environment. Brain training is based on the idea that some cognitive activities benefit brain function. Through mentally demanding challenges, the brain is kept in shape. The hope is that specifically designed exercises will strengthen key brain functions, much like the body's muscles are strengthened by lifting heavy weights.

Brain training as a science is based on the premise that some brain functions are more general and others are more specific. Those who want to get the best results from their brain training should train the most general abilities, those that are needed in a variety of challenges. In psychology research, working memory is considered the most general cognitive brain function. Working memory is the ability that allows us to hold things in mind, and use this information to solve problems. Working memory is useful in almost all everyday activities. When we cook, we use working memory to make a mental list of actions that will allow us to create a complete meal as efficiently as possible. First, we might start cooking the rice, then we cut up the fish and start frying it. While the rice and fish are cooking, we take out the vegetables and make a salad. A person with poor working memory would perhaps start with the fish instead, as it is the most important part of the meal. When the fish is ready, this inefficient cook would start cooking the rice. And so on. After an hour-long process, the cold fish dish would be served.

Working memory is used to process many different kinds of ideas and is necessary for reasoning. Therefore, it is considered an advanced intellectual capacity. People have different levels of working memory capacity. Those with the highest capacities are often those who are considered smart and score high on psychological tests that measure cognition and intelligence, but they also score high on school and college exams.

So, what does this have to do with the sense of smell? Nothing—until recently. Over the past decades, working memory researchers have produced thousands of research papers on how people process information using working memory. More than a hundred studies have investigated how working memory can be trained. The results show that working memory can be improved by training, even in older people, but that training usually only produces strong improvements in tasks similar to those trained. Many psychology and brain researchers are disappointed by this. Ideally, it would be best to find a task that could be used for effective brain training to make us better at *all* tasks that require working memory in everyday life. But the problem is that these everyday challenges are not very similar—driving a car to your destination in rush-hour traffic, remembering what to buy at the grocery store and understanding the content of the morning newspaper are quite different tasks, even though they all involve working memory. Today, few people believe that working memory training with individual tasks can help all these different abilities. But almost all researchers have examined working memory for just one type of information: visual impressions.

The sense of smell is, I believe, a promising alternative to visual working memory training. Inspired by Takako Saito's smell chess, my research team created a memory game with smells. The game is similar to a memory game where matching images must be found among upside-down cards spread out in front of the players. Instead of cards with pictures, however, we use twenty-four small tin cans with different types of tea. The task for our participants is to smell two of the tea tins during each game round. If they contain the same smell, the matching pair can be removed from the board. In a first study, we let participants play smell memory for forty days after thoroughly testing their olfactory abilities. Each daily game took about ten minutes to complete. We recruited students and other young adults with an adequate sense of smell for the study. After the training program was completed, they returned to the research laboratory to be tested again on their olfactory abilities. Their improvement was compared with that of a group trained with a standard visual memory game and that of a group of olfactory experts—wine tasters working in the Stockholm region. The results were astonishing.

The smell players showed clear improvements, even in tasks where they had not practiced. They became better at smelling nuances and detecting differences between similar but not identical smells. Even more interesting, their ability to name smells had improved, and they were now able to say which smells we presented to them, even though they were often unresponsive before. The training program was carried out in isolation, and the players did not have to name the smells at all. Nevertheless, playing the game led to clear improvement. When we compared the players' abilities with those of the wine experts, we saw that they were able to both distinguish smells and name them at the same level as the experts—after only forty days of playing.

But the most interesting result was found when we tested the memory games themselves, both visual and olfactory. Those who had played the regular visual memory game became very good at it after the training program. The same applied to those who had played the smell game—the training resulted in a clear improvement in their performance. And they also got better at the visual memory game! However, those who played the regular picture memory game did not improve at all on the smell game or any other smell task. So by replacing images with smells, we were able to promote improvements in several different tasks. Here we found the strongest evidence that brain training with smells can be just the kind of effective training that has been called for, one that leads to improvements even in tasks that are *different* from the ones trained on.

In a follow-up study, we had older people train in the same way, with smell games or visual games for forty days. The results confirmed that the older participants were affected by the training program in much the same way as the younger ones, although their improvements were slightly more modest. Thus, even in the elderly, olfactory training led to improvements. We also investigated a range of cognitive abilities in the elderly by testing visual working memory and long-term memory. The results showed that olfactory training and visual training were equally effective, but the improvements were modest regardless of the type of training. So the conclusion is that odor-based cognitive training provides some positive effects on the sense of smell, which can only be achieved through odor-based training. However, olfactory training also seems to have positive effects on visual cognitive abilities—as strong as training with visual tasks.

Smell-based brain training could thus become a new method for keeping the brain in shape in old age. So far, only a few studies have examined the biological effects on the brain, but their results are promising. A French study had participants train for three days to describe and name household smells. Before and after this short training program, the participants' brain activity was examined during the smell exercise. After the participants had learned to describe the smells, a completely different and richer pattern of activity emerged in the brain. New areas of the frontal and parietal lobes, which are far from the olfactory brain, were now engaged by the smells. Our own olfactory training study produced similar results. The older participants who trained with the smell game showed changes in their brains. Areas in the frontal and parietal lobes now showed signs of being more connected to other brain areas than before. They increasingly appeared as "hubs," nodes or connection centers for brain-wide networks. One area of the cortex of the parietal lobe may have even grown slightly thicker in those who trained with smells and learned to get better at both olfactory and visual tasks. This area is known to help us with spatial understanding, navigating and remembering where things are located in our environment. Training the ability to remember where the matching odor pairs were placed would now make this brain area more connected to the frontal lobe, which decides our movement patterns. It seems that the smells helped to create improved connections in the brain's busy information flow.

You can build your own smell memory game similar to the one we use to train the brain. All you need are pleasant smells packaged in a way that makes it impossible to tell which are related. Tea is perfect for this purpose. Packaged in their white bags, teas are hard to tell apart, but each variety has its unmistakable smell. Put the tea bags in small tea jars, glass or plastic bottles with lids to preserve their smell. Ordinary jars with spices work well too—just make sure you use identical lids and cover the sides with tape so you can't see the contents. As you practice, concentrate on the smells and think about the thoughts and feelings they evoke. How would you describe the smells? After a few weeks of getting used to the smells, swap them for new ones! The brain usually benefits from variety, and this also applies to scents.

Research on smell-based brain training is still in its infancy. But the results so far suggest that olfactory training might be a promising way to enhance the sense of smell but also broader brain functions. A 2017 German study found that elderly people who trained with a simple olfactory training program for five months—the training consisted of smelling concentrated clove, rose, lemon and eucalyptus for a few minutes every morning and evening—showed improved olfactory abilities. But other abilities were also affected. Participants scored better on a test that measures verbal ability. Perhaps most important, according to the German researchers, the olfactory training led the elderly to report reduced depressive symptoms. The results need to be repeated and confirmed by other researchers, but they are promising. If smell training can have such broad and positive effects on both cognitive abilities and well-being, it could become a popular activity in the future. In many countries, you see elderly people playing chess in parks. Will they play smelling games in the future? The coronavirus pandemic in recent years has led to an increased interest in smell training. The next chapter is about the virus that stole all the smells—and what we can do to bring them back.

CHAPTER 10

THE VIRUS THAT STOLE ALL THE SMELLS

In THE SPRING of 2020, people all over the world woke up with a new and strange experience: their morning coffee had completely lost its taste! After asking their loved ones, they realized that it was not the coffee that was off. It was their ability to smell that had disappeared. Perhaps the strangest thing about this sudden disappearance was that these people often had no explanation as to why their sense of smell had suddenly stopped working properly. They felt healthy, had no nasal congestion or sniffles that could have blocked their nose. It was a frightening feeling for many, as they realized over the next few days that all the scents of food and drink, of all the people around them, all the natural smells from the flowers, forests and lakes, were as lost as the aroma of their morning coffee. Since then, 772 million coronavirus infections have been confirmed. For a large proportion of those infected, the sense of smell was affected. How did the virus steal our smells, and what can we do about it?

The new coronavirus received attention in January 2020 after several cases were detected in the Wuhan province in China. In mid-March, stories began circulating on social media of people suddenly losing their sense of smell and taste. Then it snowballed. On March 23, two British doctors warned of a sharp increase in patients with unexplained loss of smell. This coincided with the trending of "smell loss" in Google searches.

The coronavirus hit Stockholm particularly hard in the spring of 2020. My research team's work was put on hold when we shut down the lab in mid-March. This was the beginning of a long period of isolation and working from home. We didn't know then that the virus would make the whole world aware of the importance of the sense of smell. During the coronavirus pandemic, the research community demonstrated a remarkable ability to adapt and work through the acute challenges posed by COVID-19. What started with a lockdown soon evolved into a period of intense work and new international collaborations to better understand and fight the new virus.

There were early reports that the SARS-CoV-2 coronavirus affected the sense of smell. However, this was not something that was initially given much attention. The loss of smell was initially thought to be similar to that which occurred in colds, where nasal congestion and mucus secretions help to disrupt the sense of smell. But eventually it was realized that the new virus was different. It was more aggressive and seemed to lead to an immediate, complete elimination of the sense of smell. Olfactory scientist Danielle Reed estimated in February 2022 that the COVID-19 pandemic might have left 20 million people with a permanently impaired sense of smell.

We smell and taste researchers quickly came together to meet this new challenge. On March 24, I participated in a videoconference on the new reports, along with hundreds of smell and taste researchers. My colleagues from Europe and the US created an effective organization and, just one week after the meeting, the number of participants had grown from one hundred to nearly five hundred. The team created a survey that asked participants if they had experienced symptoms of the disease in the nose and throat, and if their sensory abilities had changed as a result of the disease. Thanks to the commitment of the various participants around the world, the survey was soon translated into

thirty-five languages and published online. I did the Swedish translation together with my colleagues, and we could soon see how more
and more Swedish participants reported a loss of sense of smell.

The survey showed that the sense of smell in patients with confirmed COVID-19 deteriorated by an average of 80 percent. More
than half of the people with confirmed COVID-19 lost their sense of
smell completely, making loss of smell a particularly strong symptom
of COVID-19. Other symptoms, such as headache or fatigue, were
also present in many patients, but as these symptoms were also present in those who did not have COVID-19, but only a common cold,
they were not as good markers for the coronavirus. The sense of taste
and the trigeminal sense, which reacts to chili peppers and other hot
or pungent substances, were also severely impaired. The loss of taste
was particularly difficult to assess because people often report "losing
taste" when in fact it is the sense of smell that has been lost. Common
viruses can cause loss of smell, but rarely loss of taste. Here our survey
had an advantage because we asked about individual tastes. A full
45 percent of COVID patients reported a loss of salty taste, and this
taste is rarely confused with smell, indicating that there was a real loss
of taste.

What distinguished the loss of smell in COVID-19 from that in
common colds was that it was immediate, usually total and absent of
other nasal symptoms such as swelling and runny nose. Newspapers,
television and radio reported on the new findings. But the authorities
were slower to act. The Public Health Agency of Sweden described
loss of smell as an "unusual" symptom—even though it was the most
common. The World Health Organization (WHO) did the same,
and many countries' public health authorities downplayed the role
of the sense of smell. Perhaps it was thought that people could not
really judge whether they themselves had lost their sense of smell.
After all, it had long been described as a subjective and unreliable
sense. Perhaps because of the prejudice against the human sense of

smell, many countries' public health authorities ignored the reports, focusing instead on fever, cough and headache. But the sudden loss of smell was obvious to those affected. And I like to think I helped protect the public when I repeated to journalists in an almost parrot-like manner that readers and listeners who suddenly lose their sense of smell should isolate themselves immediately.

It was long a mystery why the COVID-19 infection hit the sense of smell so hard. But finally, researchers discovered how the virus finds its way into the body, via so-called support cells found around the olfactory receptor cells in the olfactory mucosa of the nasal cavity. The support cells keep the mucosa in order, helping to maintain the chemical balance, but they also clean the mucosa by taking care of substances that could otherwise cause damage. The support cells contain two genes that influence the shape of the outer membrane of the cells. The coronavirus has found a way to attach itself to the membrane of these cells, and from there it enters the cells and breaks them down. When the coronavirus breaks down the support cells, the consequence is that all other cells are thrown into chaos. In this eroded environment, which can now be compared to the ruins of an ancient city, new receptor cells will not always be able to find their way back to the brain. Communication is interrupted.

The total loss of smell lasted for different lengths of time for those affected. On average, there was a 50 percent recovery in just over a month, according to our study. But this figure hides a wide range. For some, it went away in a few days. For others, the sense of smell has not returned at all. The long-term consequences of COVID-19 are still not fully understood. It has proved difficult to make predictions about this new and elusive disease. However, research is ongoing about the long-term effects of the disease on the sense of smell. Staff at Danderyd Hospital in Stockholm were recruited for a large study to map the impact of COVID-19 on them. It has become one of the best Swedish data sources from the pandemic, thanks to the loyalty of the staff who, despite great hardship, contributed their time to help science. A clear lesson learned is that it is mainly those who had severe symptoms at the time of infection who have long-term problems. The patients who had severe COVID-19 had significantly more long-term symptoms after eight months, and loss of smell, anosmia, was the most common. Anosmia was present in 25 percent of those with severe COVID-19 but only in 10 percent of those with mild COVID-19. A full year into the study, 6 percent of COVID-19 patients had not yet recovered their sense of smell.

For the vast majority of people, losing the sense of smell means a significant reduction in quality of life. Depression and sadness are much more common in the group of people who have lost their sense of smell than in the general population, and more common than in those affected by blindness and other sensory impairments. The most common problem is a loss of appetite: When you do eat, you don't enjoy it—everything tastes like cardboard. However, this does not mean that people stop eating after losing their sense of smell. Some eat less, but others eat more—the loss of smell also makes it difficult to feel full. Something as simple as cooking for family or friends becomes difficult for anosmics. It's impossible to taste the food to check if you've used the right amount of spices and salt. For many people, the sense of smell is an important part of their job. For example, restaurant staff who can't answer customers' questions about the taste of their food, or preschool staff who can't sense when it's time to change diapers, find it difficult to perform their normal duties.

Many patients also describe how they lose their grip on life and have a changed relationship with the world and with themselves. "I feel alone in the world and distanced from myself." "I feel like I don't exist." "I can't smell my own odor, and that makes me feel like I am no longer me." Not being able to smell anything from your own home can make you feel out of place. Add to that the constant worry that they are smelling bad themselves, and it is easy to imagine the suffering involved.

Professor Pierre-Marie Lledo appears on the big screen. He apologizes—the Institute Pasteur has not allowed him to travel from Paris to Stockholm due to the pandemic situation. The rest of us are sitting in a hotel in central Stockholm, it's the beginning of September 2021, and just as the season teeters indecisively between summer and fall, so does the pandemic. Do we dare to relax the restrictions, how well does the vaccine protect against the new virus variants, can we travel now? It is the annual meeting of the Scandinavian Physiological Society. Those of us attending the conference are a little uncomfortable meeting in the same room; waving and scattered chairs have replaced handshakes and crowding in the coffee queue.

During the pandemic, physiologists have realized that the sense of smell plays a key role in the COVID-19 pandemic. Pierre-Marie shows his evidence for "neurotropism" in SARS-CoV-2. This means that the virus can *enter the brain—via the sense of smell*. The coronavirus can, in theory, enter the brain in three different ways. The first is through the olfactory receptors and the olfactory nerve. The second is through the vagus nerve from the lungs and gut. The third route is through the blood circulation and involves the virus crossing the blood–brain barrier into the brain. Hamsters can contract COVID-19 in the same way as humans—by attachment to cells in the olfactory mucosa. This made them particularly promising laboratory animals. They get tired, their fur is ruffled and they move more slowly. But they don't seem to die from the virus and after a few weeks all the animals are healthy again. The hamsters' sense of smell is tested with the "buried food test," which is exactly what it sounds like: food is hidden in the sand and the hamsters have to sniff it out. Healthy hamsters can do this in a few minutes, but the hamsters with COVID-19 were unable to find the food at all. The animals' olfactory mucous membranes temporarily lost their cilia, making them unable to capture the odor molecules in the air. Later, Lledo and his team found that the hamsters' olfactory receptor cells were also infected. But the most worrying finding was that the virus did not stop there, but could also be detected in the olfactory bulb. This means that the virus had entered the brain via the olfactory cells.

If the virus can enter people's brains through the sense of smell, it can do a lot of damage. The olfactory system is central to the brain, and the virus only needs to travel a few more centimeters to get to the areas that store memories, create emotional responses and connect different sensory experiences. What if these experiences were to change in the same way as the sense of smell? Imagine if your perception was so distorted by your visual brain that you wanted to vomit every time you saw people dressed in gray! Or perhaps you wouldn't be able to see bald people at all; they would become invisible to your eyes. And people or objects that you previously thought looked awful would be transformed into divine beauty, thanks to the virus. Or your memory would be wiped out so that you could no longer remember the simplest information. You would be confused by a basic conversation, unable to find your way home from work and hardly able to perform any tasks. In such a bizarre and horrifying world, our emotions, thoughts and memories have been affected by lapses and distortions, just as the sense of smell is affected by the coronavirus. If the sense of smell is a portal for the coronavirus to do damage in other parts of the brain, it would be very worrying. There is some indirect evidence to suggest this. Those affected by loss of smell after COVID-19 have been found to have a thinning of the cerebral cortex precisely in the regions that process smells and tastes—the olfactory brain has shrunk. But I believe that it's not as bad as it sounds. The shrinking of the olfactory brain is probably not a *cause* of the loss of smell, but a *consequence* of it; when brain areas are not used, they tend to become thin, much like a muscle that is not used. Scientists have different views on exactly how far the coronavirus can get into the olfactory brain. A German research team analyzed the brains of people who had died from COVID-19. They found that the virus had made its way up to the olfactory bulb at the base of the brain—but traces of the virus were found on the outside of the bulbs, not inside them. The researchers therefore believe that the virus does not enter the neurons of the olfactory brain, and there is no strong evidence to suggest that the virus enters the brain via the sense of smell and causes brain damage. Humans are not hamsters, although our olfactory organs are similar to theirs. And the most frightening theories about the ravages of the coronavirus in the brain have not been proven in humans.

Even if our brains are not infected by the coronavirus, the virus can cause long-term problems in the olfactory mucosa. Lledo and his colleagues scraped the olfactory mucosa of people infected with COVID-19 and found the virus in the olfactory receptors and surrounding cells—even though some patients had been healthy for months! Lledo believes that prolonged infection of the olfactory mucosa can be a problem for patients in the long-term. "We like to think that infections follow a certain timeline: first we are besieged by the virus, but then the body's immune system wins and we get healthy." The impact of the coronavirus on the sense of smell follows a different logic. Sometimes the body controls the virus and the sense of smell returns to normal, but then the virus can make a comeback. It reactivates in the olfactory mucosa and the sense of smell is affected again. Odor symptoms can therefore come and go, seemingly without warning. But beneath the surface, there is a long tug-of-war between the virus and the body's immune system. The idea that the virus remains active for many months is frightening. But the theory has received support from an unexpected source—smell-sensitive dogs. Since the beginning of the pandemic, another French research team has been training dogs by having them sniff the sweat of coronavirus patients. The researchers collected sweat from patients who held a cotton ball in their armpit for five minutes. These cotton balls were then introduced to the dogs. The latest discovery is that not only are the dogs able to recognize sweat from patients in the early phase of the disease, but that the dogs also react to the smell of sweat collected from people suffering from long-term symptoms, i.e., long-term COVID. The dogs' olfactory responses suggest that the virus may remain active in those affected, leading to persistent loss of smell, fatigue, coughing and stomach problems. But exactly what signals the dogs' noses are picking up is still unclear.

In the fall of 2020, I started receiving a new kind of email. Previously it had been people with impaired sense of smell who had contacted me. Now I was receiving more and more heartbreaking emails from people who had *regained* their sense of smell. But like a horror movie where a beloved relative has risen from the grave as a murderous zombie, the smells had returned in a distorted, disgusting form. This is what is commonly referred to as *parosmia*. One person described it this way: *Other symptoms disappeared quickly but it took six weeks for the smell and taste to return. When it came back, a lot of things had changed, a lot of things I usually like now taste and smell bad: coffee, chocolate, wine, berries, vanilla, all perfume, soap etc. The list goes on. Even outdoor ("fresh") air smells bad and sometimes I even have trouble drinking tap water. I have been to an ear/nose/throat specialist. He took a quick look at my throat and said that nothing was wrong and that I should just wait and hope. When I asked if I had parosmia, it seemed like he didn't know what it was.*

Parosmia, distorted and unpleasant olfactory sensations, is a strange olfactory phenomenon that affected many people during the pandemic. How parosmia occurs is still something of a mystery, and there are several theories. One of them is that parosmia does not occur in the olfactory mucosa at all, but in the brain. A Canadian research team found a patient with a lesion in the brain's insula, and the patient's sense of smell distorted all smells to be disgusting. The insula is an area that connects smells, tastes and the emotions associated with them. So it's perhaps not surprising that damage to the area can lead to distorted emotional responses. But this patient had suffered brain damage from a cancerous tumor, and did not have COVID-19. While the brain theory remains a theoretical possibility, few, if any, researchers believe that parosmia caused by COVID-19 is due to brain damage. Rather, it is due to changes in the olfactory mucosa. Meat, coffee, cucumbers, watermelon, peppers, peanuts and fried food are all foods that recur in the stories of corona patients' distorted sense of smell. While the smells are normally popular, many parosmics now find them disgusting. What the smells have in common is that they contain odor molecules to which humans are particularly sensitive. One theory is therefore that the smell of the most powerful molecules returns about six months after the initial viral infection, when our body finally wins the battle against the virus, but that this creates an "imbalance" in the olfactory brain, as all the other molecules that give food its rich and pleasant smell are still absent. Since each smell is a bouquet of odorants, this weakening can disrupt the overall picture. But that doesn't explain why the coronavirus has given rise to entirely new, and disgusting, odor experiences.

British chemist Jane Parker conducted a careful and interesting research study using gas chromatography, in which a sophisticated machine heats different substances so that their odor molecules are released at different times for chemical analysis. The machine also has a small funnel where the different odor molecules come out, one by one. Jane Parker loaded the machine with coffee, put people with parosmia in front of the machine and asked them to describe how they experienced the different smells. This allowed her to map out exactly which molecules were creating the distorted experiences.

Patients reported distorted experiences for one-third of the smells. We recognize the strongest parosmia trigger from the first chapter of the book—a type of mercaptan smell to which humans are extremely sensitive. But different patients react to different molecules, so there is no known molecule that creates the "parosmia smell" in everyone.

Coffee and chocolate drink powder have a large overlap of molecules. For poop, there are six different molecules that produce distorted sensations in parosmia, and all of these odors are actually also present in the smell of both coffee and chocolate drink powder. However, many parosmics cannot smell at all the odor that usually dominates in poop, and which most people find so disgusting—the odorant skatole. This is why, in some cases, parosmics think poop has started to smell good.

Odor distortion is debilitating and there is still no cure. Sufferers report three main ways in which parosmia causes suffering. The first is a loss of appetite, making it difficult to enjoy food and drink. In severe cases of parosmia, there may only be a few types of food that the person can eat without gagging. For some, this eventually leads to malnutrition. Thirty percent of parosmics lose weight. In a few cases, it can be life-threatening. One German woman suffered such severe parosmia from cancer treatment that she stopped eating altogether. The scent of the food caused such immense discomfort that it was impossible to get her to eat. She lost so much weight that the doctors realized she was risking her life if she didn't start eating. The solution came from an unexpected source: an ordinary nose clip. With the nose clip on, the scent of the food did not provoke such strong reactions anymore. The patient gradually started eating more and more, and the nose clip probably saved her life.

However, not everyone is affected in the same way. Parosmia can lead to unhealthy eating habits, or eating more than before. This is why 15 percent of people actually gain weight after having parosmia.

The second cause of suffering with parosmia concerns our close relationships. When the smell of our children or our partner stops making us feel good, and instead makes us nauseous, it is like living in a nightmare. Some patients tell us that their boyfriend has to sleep on the couch because they can no longer stand having him around. Sex is out of the question; smelling your partner's scent leads to rejection instead of intimacy. And that brings us to the third suffering: the anxiety and depression that come from being constantly nauseated by smells. Many patients report not being understood—and how do you explain to your partner that you can no longer stand their smell? Body odors are so emotionally charged and so personal. There is a strong sense of loneliness when you can no longer share the olfactory worlds of others.

How do you adapt to your new situation? Since odor distortion also affected people before the coronavirus pandemic, there is research to learn from. It is likely that odor distortion has been a problem for as long as humanity has been affected by cold viruses. As early as the late eighteenth century, the physician Erasmus Darwin (Charles Darwin's grandfather) described what he called *olfactus acrior*, a hallucination of the sense of smell. However, Erasmus Darwin did not explore the subject further—perhaps, like many other scientists, he thought the sense of smell was unimportant—and he advised sufferers to "stuff starch up the nasal cavity." It is unclear how Erasmus Darwin got the idea for this remedy. It should, perhaps, be clarified that it has no scientific support whatsoever. The most common strategy, according to research by Ebba Hedén Blomqvist at Karolinska Institutet, is to simply try to get used to your new situation. This is also what many doctors tell patients they cannot cure. People can certainly adapt and accept their situation—most things can be adjusted to—but it would be sad if this advice is the best available to affected patients. It is therefore encouraging to see an increasing number of people using active strategies, such as seeking more information about their condition. People who understand how their symptoms have arisen, and that they are not alone in their situation, often feel a little better. Parosmics can also advise each other on ways to make life easier. One common trick to mitigate odor distortion is to serve food ingredients in separate bowls. Another is to focus on the smells that can still be enjoyed. Cinnamon is the most popular smell among people with olfactory distortions—almost no parosmics are disgusted by the smell of cinnamon. In fact, it has become common for parosmics to replace their minty toothpaste, which causes strong feelings of disgust, with a cinnamon-flavored toothpaste!

One consolation for those affected by parosmia is that it usually goes away. But it can take time. When I started receiving emails from parosmics during the coronavirus pandemic, I contacted Steven Nordin, my former supervisor, to explore an older data set that had suddenly gained new relevance. This is the Betula study, the large research project mentioned earlier that has examined the memory and health of adults in my hometown of Umeå over several decades. We examined the questionnaire that contained queries about the sense of smell, and discovered questions about distorted smells. It turned out that 5 percent of our sample (which was a good representation of the general population) reported parosmia when the question was asked around the year 2000. This means that about half a million people in Sweden, which has about 10 million inhabitants, have olfactory distortions. The distribution of parosmics was independent of age, gender and educational level—all groups were equally likely to be affected. Nor could we see any distinctive features related to parosmia in either memory capacity or olfactory ability. The brain seemed to function normally in parosmics. The really interesting result came when we looked at how many people recovered from their parosmia. The results showed that 83 percent of parosmics had recovered before the next measurement was taken five years later. After another five years, recovery had risen to 90 percent. So one in ten did not regain their normal sense of smell within a decade. When we then traced these results back to the first measurement, we saw that one factor above others could predict whether parosmia would be long-lasting. This was if the person also reported "hallucinating" smells, i.e., sensing smells that weren't there. Then the risk of chronic parosmia was much higher. From this data, we could not say anything about how difficult the olfactory distortions were to live with for the people affected. But we compared their well-being to see if their general health was good or if they had problems with their appetite, if they were depressed or lonely, if they felt anxiety or worry—the common complaints reported by patients with olfactory distortion or loss. The results provided a more encouraging picture, as we found no deviation from normal. The parosmics either seemed to be able to find ways to live with their distorted sense of smell, or the symptoms subsided over time.

The latest coronavirus is one of many. But earlier viruses have also proved highly capable of causing anosmia and parosmia, which is why we olfactory scientists have long known these terms. Most viruses do not have well-known names; we simply call them "flu" or "cold." Even before the coronavirus pandemic, many suffered from loss of smell and distortion after viral infection. I suspect that the parosmia we found in our older data is different from that caused by the aggressive SARS-CoV-2. In one part of the Danderyd Hospital study mentioned earlier, Johan Lundström and his colleagues investigated smell changes among the hospital staff during the COVID-19 pandemic. The results show that both loss of smell and parosmia are a common consequence of COVID-19. Fifteen months after infection, the risk of loss of smell had almost doubled; 37 percent had impaired sense of smell compared to the non-infected comparison group, where 20 percent had an impaired sense of smell. However, parosmia was the most common smell-related symptom fifteen months after infection. Half of the participants infected with COVID-19 reported parosmia, but parosmia was only present in 5 percent of the non-infected comparison group, which means that the occurrence of parosmia was tenfold in those who were infected with COVID-19. Most of these had a mild degree of parosmia, and probably only a few had suffered as badly as the person who emailed me describing their symptoms. So there is hope that most parosmics can either recover or learn to live with the odor distortions. It is likely that the most severe cases of parosmia reported during the pandemic are not representative of the general population. But for those who have been severely affected by parosmia, it is of course no consolation that they have been unusually unlucky.

While it is hoped that a cure will eventually be found, medical science has unfortunately not yet found effective ways to treat loss of smell and distortion of smell. In fact, medical research has not prioritized these patients in the past. Only a few treatments have been tried and the results are not too promising. There are drugs that may affect the sense of smell. One example is gabapentin, which is commonly used to treat epilepsy. However, its effects on loss of smell are not fully understood and therefore gabapentin is not usually recommended by doctors. Another treatment is to destroy the olfactory mucosa by cutting or burning it. However, this is usually only seen as a last resort for those with very severe problems. The coronavirus pandemic has increased interest in research on how to restore the sense of smell, and now that the need is so great, hopefully research will also progress in the coming years.

One of those affected by the coronavirus pandemic is Richard Juhlin, the world's leading champagne expert. After being laid out with a fever for two weeks, he realized that what he had been so afraid of had happened: the coffee no longer had any taste. For Juhlin, a lingering smell loss could have been the death knell for his career. But Juhlin is a trained physical education teacher, so he knew how well the body can recover from injuries with the right physical therapy. He started retraining his sense of smell using a smell exercise. Discerning the subtle nuances of exclusive champagnes was no longer possible. Juhlin's salvation was the curry jar in the pantry. Curry is a complex mixture of smells with elements that he recognized well. He knew exactly which nuances seemed to be missing from the mixture. The exercise was as much intellectual as sensory; he imagined the nuances that should be there, and sniffed the curry jar. Fortunately, after a few days they appeared, one by one, like old friends at a reunion, and he recovered his smell acuity.

Research supports olfactory training as a way to help rehabilitate the sense of smell. For those who have lost their sense of smell after COVID-19 or another viral infection and do not regain it within two weeks, olfactory training is the only method recommended by the world's leading experts. The knowledge of olfactory training did not start with COVID-19, but in the 1990s, when scientists started to take an interest in how the brain was affected by different types of stimulation. Brain researchers used rats and mice and provided their cages with a variety of smells. The effect was clear—the olfactory brain was positively affected by the smells. The stimulation led to an increase in the activity of the olfactory receptor cells, and for rodents that are presented with different odors on a daily basis, this stimulation leads to an increase in the lifespan of the cells. The sense of smell depends on a constant supply of cells that can respond to the smells around us. The animals that sniff their way through life in this way have an improved regeneration of nerve cells in the olfactory mucosa. The new cells come loaded with fresh olfactory receptors that increase sensitivity to environmental odors.

A growing body of research suggests that olfactory training may work in a similar way for humans as it does in rodents. In 2009, Thomas Hummel's research team at the Technical University in Dresden published the results of the first study in which people with loss of smell were prescribed olfactory training. The patients were asked to sniff lemon, eucalyptus, rose and clove for five minutes every morning and evening over a twelve-week period. Hummel and his colleagues found that the patients who received smell training had a better recovery than those who did not; 28 percent had an improved sense of smell after smell training, compared to only 6 percent of the control group.

Since Hummel's first study on olfactory training in patients with loss of smell, the results have been repeated. A 2021 compilation shows that sixteen studies have calculated the effect of training the sense of smell. The results have confirmed the conclusion that olfactory training leads to better prospects of recovering the sense of smell. To be precise: the odds of recovery are 2.77 times better for those who do smell training. And for those whose sense of smell was damaged during a viral infection, the chances are better than for those who have hit their head—another common reason for loss of smell. Unfortunately, recovery does not work for everyone, and it is rarely 100 percent. Most people who have long-term damage usually have to live with a weak sense of smell even if they train. But there is a big difference between having a sense of smell—even if it is weak—and having none at all. The smell of the food we eat, our partner's hair shampoo and maybe even our babies' diapers create some of the important feelings we need to feel good.

We have created a website, www.smelltrainingapp.com, that is used to treat people diagnosed with loss of smell at the Karolinska University Hospital in Sweden. Patients complete traditional olfactory training using cinnamon, olive oil, thyme, coffee or other smells they have at home. On the website, which is free to use and available in English, German, Spanish and Swedish, patients enter information about how they experience the smells, day by day. This helps us to answer important questions: How long does it take for an odor to become stronger, and which patients recover best? Does it matter how much you exercise? There is still a lot we don't know about the recovery process, but this information gives us a better idea of what happens during the months when patients regain their longed-for sense of smell.

Perhaps you are one of those who have been affected by a rapidly deteriorating or distorted sense of smell. Perhaps you are worried that your sense of smell will remain altered and not return to normal. I want to give you some advice that hopefully can help you maximize the chance of recovery and live as full a life as possible.

My first piece of advice to you is to see a physician specializing in the ear, nose and throat (ENT) area. Ask them for an evaluation and to have you undergo a smell test—such tests are administered at some, but not all, ENT clinics. The advantage of taking such a test is that you can have your symptoms confirmed and explained by an expert. People with changes in their sense of smell not only suffer from their symptoms but also from the lack of understanding from those around them—their troubles are not visible from the outside, and others often do not understand their suffering unless they themselves have experienced it. The encounter with a physician can, therefore, have a positive effect beyond the strictly medical diagnosis and treatment, as it validates the patient's experience. Follow the doctor's advice, but do not go there with expectations of rapid improvement. Unfortunately, the doctor's ability to help you is limited; at the time of this writing there is still no method that can help everyone toward a complete recovery.

It is too common for patients to feel disappointed after being evaluated by their doctor. While it feels validating to have the symptoms confirmed by a specialist, after the visit the patient is often left with the advice to try to adapt to their condition and without receiving any new tools to manage it. My second piece of advice is, therefore, to seek support from others experiencing the same thing. On the internet, there are many forums for anosmics and parosmics. I collaborated with AbScent, an organization based in England. On its website and in social media groups, members shared practical tips and research-based knowledge about olfactory changes. In 2024, AbScent's founder Chrissi Kelly discontinued AbScent but started a new initiative called Chrissi Kelly on Smell. Support and advice from others who are in the same situation can be indispensable. The community can be helpful in many ways, supporting changes in eating habits, emotional and relationship problems and other aspects of life affected by altered sense of smell. Patients gain strength from not standing alone with their troubles, and there is a sense of shared optimism and determination.

My third piece of advice is to engage in smell training. There is now scientific support for the idea that smell training improves recovery from some forms of impaired or distorted sense of smell. Unfortunately, not all groups benefit equally from smell training. Among the various causes of smell loss, viral infections, such as from the coronavirus, seem to offer the greatest opportunity for recovery with the help of smell training. When smell loss is caused by a head injury or a congenital condition, though, smell training has no clear effects.

The sooner training begins, the better the chance of recovery. Set aside five to ten minutes for smell training every morning and evening. Choose a few smells that you like and that have previously given you familiar, distinct odor sensations. The most common smells used in research studies are lemon, eucalyptus, rose and clove. But there is no reason to believe that only these smells are effective; other smells should work just as well. Using more than four smells and changing smells about every two weeks can be beneficial; at least, variation may increase your motivation for what otherwise may seem like a somewhat boring task.

Smell training should be done twice daily for at least four months. It sounds like a long time, and for those who are reminded of your smell loss every day when you sit in front of your smell jars, it can be frustrating. You will surely want to give up at some point, especially if you do not experience benefits right away. But consider that the training may yield results that you are unaware of. We are rarely aware of the biological processes happening within us, and we do not observe the activities of the millions of olfactory cells recycling in our olfactory system. Try to motivate yourself, perhaps by thinking of these cells rebuilding in your nervous system during the training period. If your sense of smell improves, there is still a risk that motivation will decline. Many people think, Now that my sense of smell has gotten a bit better, can't I let go of the smell training? However, those who continue smell training will generally experience an even stronger recovery than those who prematurely end their training.

The research on odor training is far from complete and many important questions remain to be answered. First, what is the actual effect of odor training? It is actually difficult to estimate with precision. Many patients drop out of research studies. They get tired of smelling blindly and do not notice any immediate effect. These dropouts, who are not always included in the research reports, create uncertainty about the results of the research, so the figure of 2.77 times better recovery may be revised as we learn more. Another unanswered question is whether there are even better ways to train the sense of smell than what researchers have tried so far. Perhaps we have only scratched the surface of the improvement potential of the sense of smell. Our brains have an amazing capacity, and we can imagine smells just by hearing their names. The patterns of activity in the olfactory brain bring these sensations to life. Can such methods be used for rehabilitation?

One of my main tasks over the last few years has been to try to develop effective and fun ways to train the sense of smell. The smell memory game was the first of our new smell methods. I have been working with game researcher Simon Niedenthal to help COVID patients stay motivated. Together we created an olfactory computer game called Exerscent. Smells are not usually associated with computer games, because computers cannot present smells. We have therefore built our own "game console." The game contains fifty-four small bottles with odorants that are usually used to train wine experts, and a small "game box" that connects to the computer. The scent bottles look the same, but on their underside there are small transmitters that register when they are placed on the game box. This allows the computer to know what you smell and ask smart questions about that particular smell. The game is simple. The computer asks what the smell is and gives six different options; you get points for each correct answer and the points are added up during the game. It is more difficult than you might think. Even those who think they have a normal sense of smell usually get a third of the smells wrong. They often confuse different fruits or flowers with each other. Peach can be difficult to distinguish from apple, and lilac can be mistaken for lily of the valley. But since Exerscent reveals the right answer, most people with a normal sense of smell can learn to become olfactory experts. After a few weeks of daily play, a normally sensitive nose can develop almost perfect accuracy. The most important result is that participants find it fun. The game engages large parts of the brain. It is challenging and provides rewards in the form of points. The player gets clear information on how their sense of smell develops, week by week, and can compare their recovery with that of others.

So there is hope that new olfactory methods can help those suffering from a lost or distorted sense of smell. But the applications of smell games go beyond rehabilitation to include users who want to take a perfectly normal sense of smell to an even higher level. The training methods could be used by wine tasters, perfumers, chefs and food company product developers. For these, the cognitive perspective of the sense of smell is ever present. The interaction of the sense of smell with our expectations, knowledge and the impressions of the other senses determines whether their new products will be an international success or a failure.

CHAPTER 11

BECOMING NOSEWISE

I DON'T REALLY KNOW what I'm doing here . . ." José Avillez takes the stage, dressed casually in a wrinkled linen shirt, jeans and sneakers. He is modest, but hardly shy. Portugal's most famous chef has the quiet confidence that makes the audience—two hundred brain scientists and psychologists gathered at a smell and taste research meeting outside of Lisbon—listen intently to every word. As head chef of Belcanto, the Michelin-starred restaurant in Lisbon, Avillez has a lot to say about smells and tastes, but also about psychology and the brain. In a chain of ten luxury restaurants, Avillez and his colleagues develop around two hundred new recipes a year. You might think that the chefs and head chefs are tasting their way to new recipes that will cause a sensation on the menus. But that's not how it works at all. For Avillez, and other master chefs, the most important tool is the "memory of the palate." Recipes are not tested in the kitchen, but in the chefs' brains. There are thousands of smells and tastes stored there, and all the associations they can evoke. Master chefs create recipes much like a writer creates a novel, or a composer creates music. They have a sense of what fits and combine different ingredients in their minds. It is both an intellectual and a sensory process. The cognitive perspective is clear to them. But the cognitive approach to food is not just something that experts master, it also shapes our experience of food and drink.

When you eat and drink, expectations are built up by what you see and hear. When the sensory impressions from the eyes and ears are combined with smells and tastes, they create experiences of either coherence and harmony, or deviation and surprise. This is the principle that governs our brains when we eat and drink. A principle that I have studied for twenty years in the research laboratory. And which, I realize, has its direct counterpart in experimental food culture, where harmony and surprise interact. I feel a kind of kinship with chefs who work like scientists, and am fascinated by how they create meals that engage all our senses. When we eat, our expectations take center stage. We have a need to confirm those expectations. The sense of smell has the characteristic of preferring familiar to unfamiliar smells, a survival principle that has kept us mammals alive throughout evolution. This is why we love "comfort food." Food that tastes, smells and looks just like it did when we were children, and that evokes familiar feelings and memories, has a special appeal. The familiar-taste experience is a multibillion-dollar industry. The business concept of fast food restaurants is not only about fast and cheap food, but also about *predictable* food. Anyone who regularly visits a fast food restaurant is unlikely to be disappointed. The food will taste exactly the same as it did last time. That's the point. Hand on heart, haven't you, while visiting a new country with a different food culture, slipped into McDonald's and devoured a Big Mac with fries—and guilt—on the side? Don't feel guilty about it. We need the familiar tastes and smells that give us a sense of security in new environments. Predictability is the goal of fast food restaurants. It is their key to success, but also their limitation. Because you will not be surprised in a fast food restaurant. And you definitely won't learn anything new and exciting. On the other hand, for those who want to have new food experiences and explore food culture in a deeper way, contrasts and new, innovative combinations are attractive. Experimental cooking deepens our psychological understanding of food and drink. Those who visit an experimental luxury restaurant *expect to be surprised.*

Our preferred odor and flavor combinations are largely a matter of cultural practice and habit. Take the example of eating vanilla ice cream with ketchup, a thought that disgusts most people. But why? There is nothing stopping us from eating it. Our genes have hardly instructed us that these two flavors would be disgusting in combination, because both ketchup and vanilla ice cream are cultural novelties. And almost everyone likes them—separately. Yet the combination is so disgusting that it's included in researchers' disgust sensitivity questionnaires. Perhaps you've found that context can make an unexpected smell unappetizing, even though it would otherwise be a favorite. Perhaps at some time you'd been given coffee when you were expecting tea or white wine when you expected water. It can be an almost shocking experience when the "wrong" drink lands in your mouth. Our expectations, and the context in which they are created, are the most important ingredient in food and drink.

Master chef Avillez tries to surprise people and create taste sensations that make them think. One example is how he works with fruits. He uses strawberries to create pleasant associations with childhood. He then breaks up this sense of safety by giving the strawberries a grilled taste, or by making pickled or unripe strawberries a tart accompaniment to meat or fish. A master chef also doesn't shy away from even bolder experiments. Avillez once created a cake made with an animal's blood. The unsuspecting guests ate with great appetite, but when they learned that the cake contained blood, their first reaction was to spit it out. But the chef explained that blood appears in a variety of dishes, such as blood sausage, which is a traditional dish in Portugal. After listening to the chef's story, which provided a background to the taste experiment, the guests dared to give the blood cake a second chance. The cultural associations and memories evoked by the story led to a psychological transformation. The map of expectations was redrawn in the minds of the diners. The interpretation of the taste and smell of the cake was upgraded from disgusting to fascinating. Could a chef with an interesting enough story even make us like vanilla ice cream with ketchup? I think so. The link between vanilla and ketchup has already been studied by scientists. In Germany, it was common for many years to flavor infant formula with vanilla. Researchers in Frankfurt later investigated whether this vanilla flavor could have an impact later in life. They let their study participants taste two different types of ketchup, without telling them that one of the varieties had a faint vanilla aroma. And there was indeed an impact. Among the participants who said they had been breastfed as children, only 29 percent liked the vanilla ketchup best. But in the group that received vanilla-scented formula, 67 percent liked the vanilla ketchup best. They carried with them a preference for vanilla scent that was established in childhood. This is perhaps the clearest example of how we are guided by scent memories that we don't even know we have. All food has a history. It relates to our subjective memories, but also to a shared cultural heritage—our traditions and food culture. Which food we like and which food we hate depends on how our brains interpret these legacies. Context is crucial.

People all over the world love traditional food: food that has been prepared and served in the region, in exactly the same way, for generations. For me, it's surströmming. For other Swedes, it might be lutefisk, palt (a potato dumpling with meat) or pickled herring. These are foods that evoke memories and create a sense of belonging, but whose appeal is not so easy to explain to the uninitiated. The familiar tastes and smells give us a sense of security. But it sometimes seems that we are a bit ashamed of this feeling, as if it is not a good enough reason to eat traditional dishes. Traditional recipes are therefore often described as having some ancient flavor secret that makes them irresistible. Or the food is described as having a special health effect that people in the past knew about but has now been forgotten. More often than not, such stories are just fabrications. The truth is that it is our brains, their capacity for memories, emotions and associations, that make traditional food so satisfying. The traditional recipes were often based on necessity—ingredients were added or removed to make the food sustainable or to mask unwanted flavors. The Portuguese national dish bacalhau—cod dried, boiled and salted—was widely available, easily dried and could be preserved in the cellar even without a refrigerator, making it popular with the masses. In many cultures, dried fish has similarly survived throughout history in traditional dishes, although it can hardly be described as easy to enjoy. Surströmming is produced through an acidification process that gives it its distinctive smell and taste. This process was first used in Sweden in the sixteenth century as a way to preserve the fish without using too much salt, which was expensive at the time. Thus, fermentation was used as a way to prevent the fish from rotting. Similar methods exist all over the world and both German sauerkraut and Korean kimchi are examples of fermented foods that have become popular outside of their countries of origin. Not only fish and vegetables but also traditional baked goods have been shaped by practicality. In Portugal, tourists line up to buy *pastel de nata*, the small pastry with its characteristic cinnamon flavored egg custard. Like most local specialties, this pastry has its own cultural history. The monks created the famous pastries in their monasteries, and their recipes were influenced as much by practical considerations as by taste. Monasteries produced wine on a large scale, using high quantities of egg whites in the process. The monks therefore had just as many egg yolks to use. This resulted in pastries with enormous amounts of egg yolk, and sugar to preserve them longer. The recipe has evolved over time, much like that for Swedish gingerbread cookies, which in the fourteenth century contained plenty of pepper as well as cedar oil, which would hardly suit today's taste buds.

The realization that traditional recipes were often based on necessity, rather than taste considerations, is liberating. Nowadays, we don't have to think about necessity. Skilled chefs, just like artists—or, for that matter, scientists—move with one foot in the familiar landscape of the past, and the other looking for new, solid ground to stand on.

Good food goes hand in hand with good wine. Wine connoisseurs have long been the epitome of cultivation, worldliness and often snobbery. Recently, however, wine "expertise" has become an increasingly controversial notion. Many argue that wine experts are simply frauds. Writers almost seem to be competing to bash anyone who thinks they can judge wine better than the average person. "I am going to stand up and shout: Wine reviews are bullshit," wrote Joe Power in Another Wine Blog. And Robbie Gonzalez, a writer for Gizmodo, wrote that those who are seduced by lofty wine reviews are nothing more than "pompous assholes." So it's time we used the cognitive perspective to examine what wine experts actually do, and why their opinions are sometimes dismissed as fabrications. In simple terms, the sommelier's expertise consists of three types of skills: (1) knowledge of wines, grapes, soils and producers; (2) sensory sensitivity; being able to correctly describe wines in a blind test based on this knowledge; (3) practical ability to handle the wines and their customers. The focus here is on the first two areas of competence—knowledge and sensory ability, and that is what I refer to here as wine expertise. But, as you will see, this skill involves much more than just sensitive noses and taste buds. It is, in fact, a type of intelligence.

The hatred of wine experts is actually an old tradition, as old as the word *sommelier*. The word originally comes from the French word for a packing bag strapped to donkeys and other pack animals, indicating that wine tasting was not always a particularly prestigious activity. Often it was failed cooks who were sent to the wine cellar. There, gradually, and probably driven by a great deal of bitterness and revenge, they began to create their systems for describing wines. It has evolved over the years into a vast world of olfactory and tasting descriptions, competitions, methods, literature and training.

I want to defend the sommeliers and wine experts because I know how difficult it is to name smells and tastes. Anyone who has tried to describe a wine or a perfume understands that it is a demanding task. Criticism often focuses on the elaborate descriptions of wine writers, which are perceived as being plucked out of thin air rather than objective descriptions of smells and tastes. The research also shows that wine reviews have actually become increasingly verbose, in some cases outlandish, probably because readers find the descriptions entertaining. But it is unfair to dismiss all wine expertise based on wine reviews, because the writers' job is not to be objective. They are writers, and their job is to generate interest, not only in the wine but perhaps, above all, in their own writing. The most popular wine writer is not the one who describes the wines in the most objective way, but the one who finds the most original interpretations, the wittiest formulations and the oddest associations. Anyone can start a blog and call themselves a wine writer, so the quality of analysis will obviously vary. To condemn the entire wine tasting industry because some reviews are not "objective" misses the point. Although some writers may take too many liberties, the serious wine writers show impressive knowledge and sensory ability. But even they use their creative potential, finding more and more associations with the original odor and flavor note. What wine writers do is *create meaning*, even at their most subjective. They help to give the wine consumer a richer experience. The real expertise of wine experts is not primarily reflected in the online wine reviews, but in their training, methods, and their performances in blind tests.

Earlier in history, wine writers did not even try to capture the smell and taste of wine. The ancient thinkers, such as Aristotle and Plato, believed that the human sense of smell was inadequate, and that it existed only to distinguish between harmful and beneficial odors.

Ancient Greek and Roman writers sometimes wrote about wine, but not primarily about its sensory properties. The wine writers of the time, Athenaeus, Horace and Pliny the Elder, were more practically oriented. Was the wine good or bad? Did it make you feel full? Did it give you a hangover or gas in your stomach? Only in the eighteenth and nineteenth centuries did wine reviews change. This was linked to improvements in wine production. Flavor could be controlled and became more reliable. There was also a growing market for wine. People moved to the cities and restaurants became increasingly numerous. Now it became more important to be able to assess and describe the qualities of wine. But it was still mostly about subjective experiences. Writers described wines with characteristics as if they were people. For example, in his 1920 work *Notes on a Cellar-Book*, the English critic and wine connoisseur George Saintsbury describes his favorite wine as "manly." It was only in the 1970s that the language of wine became more scientific and focused on sensory qualities. The change was influenced by the researcher Ann Noble and her "wine aroma wheel." Now an emeritus professor at the University of California, Davis, Ann Noble spent decades researching how the chemistry of wine affects its smell and taste. Her wine aroma wheel created a reference point for assessing the sensory properties of wines. Noble surveyed professional wine tasters, but also her own students, to gather concrete terms. General words like *fragrant* were eliminated as they were considered too vague. The result was a wine aroma wheel that contains both broad categories and more specific descriptions. Some categories can be understood even by a novice—nutty, fruity, floral wine smells are easily recognizable, but other categories are more obscure. What about chemical, yeasty or earthy smells? For a wine expert, odd smells such as mold or gasoline are not necessarily unpleasant, but just an objective way to describe all the nuances of the wine.

Becoming a certified wine expert is hard work. The methods used to become a wine expert are rigorously tested and arduous to master. In her book *Cork Dork,* journalist Bianca Bosker describes the extreme methods that top wine experts use to enhance their senses. They often abstain from strong smells, including coffee and spicy foods, all to keep the nose and taste buds alert and fresh. The Court of Master Sommeliers, one of the leading sommelier organizations, issues diplomas to those who pass their rigorous test. Only 274 people have achieved the highest level, master sommelier, in the whole world. Those who aspire to such an appointment can expect to invest many years of their lives and a small fortune. Ten years of work experience in the wine industry and several diplomas from the organization are required to even be considered for the tough test, which only a few pass. The organization recommends that wine be evaluated in five stages, each based on different sensations. The first stage determines how the wine looks in the glass. The second stage involves using the nose. The taster lowers their nose into the glass, closes their eyes and takes a deep sniff. In the third step, the taster lets the wine swirl around in the mouth to leave an impression on the taste buds. The experienced taster opens their mouth and sucks in air before swallowing—a risky method that, for a novice, can lead to embarrassing drooling accidents. But it's a way of allowing the aromas to travel through the pharynx and out through the nasal cavity—as you now know, these retronasal olfactory impressions are important in enabling us to give more abstract and generalized descriptions of what we're tasting. This is why champagne expert Richard Juhlin slurps so loudly when drinking his exclusive champagnes. It puts new parts of the brain to work. In the fourth stage, the wine taster assesses how the wine feels after swallowing, and how long the smell and taste remain. The final stage is a more reflective assessment. Here, all the different impressions made during the process are weighed, resulting in a final evaluation. The wine tasters do this together. The sense of smell differs from person to person, and it is also not easily captured in words. Individual discrepancies are ironed out in group discussions, and the collective evaluation becomes more universal. This is why wine experts' training rooms resemble political forums, where different interpretations are debated. "Taste like a detective, argue like a lawyer" is the motto.

The olfactory analysis is the most important step in the process. Wine tasters pick out the different smells—known as *notes* or *objects*—such as apricot, tar, sherry, olives, banana. They then characterize each object according to its relative strength and its more specific characteristics. Which olfactory note dominates? How ripe is the banana? Does the olive smell like green or black olives? Next come the mouth sensations, roughness, alcohol content, sweetness and acidity. But the real challenge is what comes next. This is when all the sensory impressions are matched against the gigantic database that the wine experts have built up over years of study. It's not just about identifying the grape variety and which vintage. They are also expected to say which country the wine comes from, which region within the country and which district within the region. All this is based solely on the senses and the memories they evoke. The wine expert's memory cathedral houses a deep knowledge of the wines, their traditions and history, the different soils and climates of the wine-producing countries, and all possible methods of production and storage. The secret of wine experts is that they don't just use their noses and taste buds. Evaluation is an intellectual process where knowledge is interwoven with the impressions of all the senses. The intelligence of the sense of smell is really put to the test when the wine experts use their detective noses. A smoky note indicates that the producer used new oak barrels—French, not American! Mint and black currant reveal the Cabernet grape. A peachy smell means Chardonnay from colder climates, while a melon smell points to warmer countries. Wine tasters are nosy *theorists*.

Yet, despite these sophisticated methods, wine tasters have difficulty gaining public respect. Perhaps you have heard about research showing that supposed wine experts can be fooled into thinking that white wine is actually red, just by adding a little colorant. This story comes from a report published by French researchers in 2001. The study has led to one of the most persistent myths about the sense of smell: that wine experts are just bluffing. Let's review this study.

The researchers recruited fifty-four students from the University of Bourdeaux's wine tasting course. In the first session, they were asked to smell and describe a red and a white wine. The following week, the students came back to the laboratory. They were again given two wines to smell, one white and one red. Their task was now to match each of last week's descriptions with one of the two new wines. The researchers found that most of the descriptions made of the red wine were now matched to the red wine. But the researchers tricked their participants—the "red" wine was actually a white wine that had been dyed red just before the smell test! And the students more often matched the odor descriptions for red wine with the wine that looked red.

Indeed, it is easy to sympathize with the students in this survey. For one thing, they were hardly experts—more like trainees. Second, the experiment was rigged. The two wines smelled similar, but the words describing red wines had to be attached to one of the two wines, so most students chose the wine that looked red. What else would they have done—say that the white wine smelled of black currant, plum, chocolate or any of the other red wine notes? This experiment has been used for twenty years now as proof that wine experts don't know what they are talking about. For example, blogger Robbie Gonzalez wrote that "not one of the 54 experts surveyed noticed that it was actually a white wine"—but this is a misinterpretation of the experiment. Perhaps the students knew very well that the wines had a similar smell; recent research has shown that experts are actually very accurate in distinguishing white from red wines and can often even deduce the exact grape variety by smell. But in this experiment they were not asked about this.

The experiment did not at all show that wine experts are frauds. But it showed something else even more important, which you are well aware of by now. The sense of smell works by taking cues from other senses. This is a characteristic that does not disappear, but *increases* with increased wine knowledge. The wine expert's interpretations *should* be influenced by all relevant sensory impressions—that's their job! The expertise lies in how these impressions are used—the color of the wine is transformed in their brains into predictions of what the wine *should* smell and taste like. This is how they can find all the different olfactory notes. The notes are not something that only affect the nose, they exist as hypotheses in the wine taster's brain long before the molecules reach the nose. The wine experts are a good example of how the cognitive perspective is necessary to understand the intelligence of the sense of smell. It is high time that we dispel the myth that wine experts are just bluffing.

The language used by wine experts is a key to their olfactory worlds. Researchers have compiled and analyzed descriptions in thousands of wine reviews. The results show that wine experts use an olfactory language characterized by colors. They mainly use two types of words: those that describe red or dark objects, such as red and black currants, violets and cherries; and those that refer to yellow or transparent objects, such as honey, butter and lemon. None of these objects appear as ingredients in either white or red wine. Nevertheless, they influence how we perceive wine smells. Thus, color is the most important factor when describing wine smells. Smells are usually difficult to describe in words. Using color-related words in this way is a clever system for the brain to create more order when communicating about our olfactory impressions. No wonder the experts are influenced by whether the wine is red or white! Instead of seeing it as a handicap if we are affected by the added color of the wine, maybe we should see it as a talent to be affected by colors when communicating about the smells of wines.

As you now know from the previous chapters of this book, our sense of smell is particularly influenced by other sensory impressions—and the cognitive perspective helps us understand how this influence works. Psychological research on food and drink shows that the senses of smell and taste can be easily manipulated by the information given to the taster or the environment in which the food is tasted. Most often, this influence is completely unconscious. Yet it can have extreme consequences. Researchers at Lund University set up a tasting table at a local supermarket. Customers were given two types of jam, blueberry and blackberry—jams that looked the same but smelled and tasted different. The researchers asked the customers which jam they liked best and recorded their answers. They were then asked to try their favorite jam again and describe why they liked it. Customers gave descriptions of why they had chosen this particular jam: it had more flavor, was sweeter and so on. The problem was that they hadn't described the favorite jam at all, but the other one! The researchers had used a simple trick to sneak in the other jam without telling customers. Less than a third of customers noticed that the jam had been replaced. Even when the flavors were completely different—apple and cinnamon were switched to bitter grapefruit— less than half of the participants noticed.

The experiment is an extreme example of how our senses can trick us. The sense of smell is particularly susceptible. If someone asks us if the food was good at the new restaurant, our response will be influenced by a variety of other experiences—how nice the waiter was, whether the noise level was high; all the senses interact, and our experience of the food is influenced. Even the cutlery matters, as was shown by Oxford University researcher Charles Spence. Too light cutlery makes food seem less tasty, and plastic cups dampen the taste of what we drink. This is one reason why airlines often use steel cutlery and drinking glasses when lighter alternatives would be easier and cheaper to take into the air.

Color is perhaps the most common flavor enhancer. A glass of strawberry juice with a strong red color is considered stronger in taste than a glass with a faint pink hue—even if the red color is added with a dye that has no smell or taste. The eye tricks the nose and tongue. Colorful food is therefore a method for increasing appetite, which is particularly important in the elderly. This is an important insight for anyone working in nursing homes and health care settings where patients' appetites may be poor. Using colorful food can reduce the risk of malnutrition.

Our sense of smell and taste is not only influenced by the color of the food—other sensory impressions also play a role. Research shows that colorful plates lead to improved taste and increased consumption. But this is mainly true for foods that are dull and monotonous in color, such as porridge, which can easily be spruced up with a colorful plate. For a colorful Greek salad, you might as well keep your white plates. For retail products, packaging is the easiest way to achieve a reinforcing effect. Unlike the store environment and staff interaction, packaging is entirely under the control of the producer. Both the packaging and the product itself usually involve the same color as the object whose flavor is added to the product. Lemon-flavored candy or soft drinks are yellow, orange-flavored are orange, apple-flavored are red or green, and so on. The reason is not just consumer education—colors help your brain interpret smells and tastes, and research has shown that the bright colors on the packaging actually make us perceive the smell and taste of the product as stronger and better. This, of course, is exploited by companies who want to satisfy our senses and turn us into loyal customers.

Have you ever wondered why chip bags are so loud when you handle them? The loud bags are not a coincidence, but a consequence of our senses interacting with each other. Research has shown that the sounds we make while eating play a bigger role in our perception of the experience than we might think. The sound when we eat crispy food travels to our jaws and creates that fresh feeling. Without this sound, we would consider chips, salad and other crunchy foods to be old and bland. In one notable study, Charles Spence and Max Zampini had participants eat Pringles chips while listening to an amplified live recording of the sound of their eating. The result was that the participants thought the chips tasted fresher and crunchier—even though the chips tasted exactly the same as usual. The results point to the important role of sound in our perception of taste and smell. However, enhancing the sound of crispy food is easier said than done. It is often both difficult and expensive, especially if the product, like chips or other snacks, has already been carefully developed to achieve maximum crunchiness. Any change to an existing product risks affecting it in unwanted ways. This is why food producers try to reinforce feelings of crispness by making the *packaging* extra crunchy—it makes us think the contents taste better!

Because the senses of smell and taste are so susceptible to influence, this leads to good opportunities for advertising agencies' messages. Sales messages can become effective flavor enhancers. "Organic eggs directly from Inger at Ostnäs farm" somehow "tastes" better than anonymously packaged eggs—the message brings to mind a well-managed, small-scale farm with good animal husbandry and all the positive feelings that evokes. However, when organic or free-range food products are compared to other food in blind taste tests, there is most often no difference—the expectations and emotions themselves enhance the smell and taste. Marketing messages have measurable effects, which can be seen in the subjective assessments of tasters. But they can also be read directly in brain activity patterns. The brain's reward area is activated when our sensory impressions create desire, liking or feelings of pleasure. The more we like something, the higher the activity. Brain research has proven that smells and tastes can indeed be enhanced by marketing. The reward system is more strongly activated when we taste a wine that we think is expensive and exclusive, compared to when we taste the same wine with a modest price tag. Similarly, psychological experiments have shown that when we *expect* to taste a strongly bitter liquid, the brain's taste area reacts strongly to the liquid, even if the experimenter in fact presented a taster with a weaker solution.

Unfortunately, marketing can sometimes have the opposite effect. Messages that a certain food is low-fat, low-calorie, sugar-free or climate-friendly can actually make consumers think it tastes worse! This is because such messages create an expectation of an inferior taste experience—and this bias is enough to shape the experience, even if in some cases it is unconscious. This is one of the reasons why vegetarian food often has a hard time catching on with meat eaters. The healthy message on the packaging hardly makes the mouth water, but such desires determine most people's purchases. Recently, producers have understood this issue and developed vegetarian options that taste great and are marketed with taste—not health or climate—as the main selling point.

But there are of course limits to how much expectations and environmental effects can alter our perceptions of smell and taste. If the discrepancy is too great, our brains will pick up on the scam. No one can convince you that a glass of completely flat soda comes from a freshly opened bottle. The flavor changes too much when the carbonation disappears. Environmental cues must be in line with the smell and taste impressions to reinforce them. This is why colors are the most common enhancement method, as mentioned, and it is often cheap and easy to color a product like the fruit it is supposed to taste like.

So the sense of smell collects our knowledge, feelings and impressions from our other senses and uses this information to interpret what we eat and drink. Yet there is a risk when producers change the taste of an existing product, not because consumers tend to notice when the taste or smell changes, but because they tend to dislike it when they learn that their favorite product is changing its recipe! This sometimes gets in the way of product development. If a company could create a chocolate bar with less sugar or fat without changing the taste, it would be a healthier and better option. However, if customers were made aware of the new recipe, they would probably still think that the taste had changed for the worse. So sometimes it's best to sneak in changes. Coca-Cola has changed its recipe many times since the soft drink was launched in 1886, but this was kept secret and consumers did not react. But in 1985 they created New Coke, a new recipe that, according to extensive blind tests, tasted *better* than the existing product. Another advantage, company executives thought, was that New Coke tasted more like Pepsi, the competitor that grew in popularity in the 1980s.

The results were devastating. Coca-Cola's customer service received eight thousand complaints—per day. Traditional Coke lovers took to the streets and ostentatiously poured New Coke into the gutter. One customer sued the company. Even Cuba's dictator Fidel Castro, a major consumer of Coca-Cola, complained about the new flavor. He described it as an example of "the decadence of capitalism." And sales plummeted. Coca-Cola was soon forced to revive the traditional recipe, now called Coca-Cola Classic, to win back traditional customers. The mistake has become a classic example taught to students in business schools. Coca-Cola learned the lesson that their success is as much about relating to expectations, memories and emotions as it is about inventing tasty soft drinks. They realized the importance of the cognitive perspective.

High-end restaurants and food producers are now increasingly turning to psychological research to enhance their customers' experiences. In his book *Gastrophysics,* Charles Spence describes how plates, cutlery, music, furniture—everything is aligned to create the right feeling. Some luxury restaurants go one step further: they want to use your individual olfactory psychology to create a unique and unforgettable dining experience. For example, if you book a table at The Fat Duck in the English village of Bray, you may receive emails asking for personal information. They want to know when and where you were born, and about your childhood. The restaurant uses this information to prepare personalized meals based on your positive feelings and memories of your childhood. Nostalgia is thus one of the latest innovations in the restaurant industry. However, The Fat Duck is no ordinary restaurant, but one of the best in the world. The restaurant's famous tasting menu costs up to USD 500 per person. So it may be a while before the restaurant around the corner brings your memories back like Marcel Proust's madeleine cake dipped in lime-blossom tea. But if this concept is successful, perhaps more restaurants will serve "Proustian" meals in the future.

The senses talk to each other and this exchange can lead us to believe that food tastes better. Those who understand the cognitive perspective can also use their insights to trick their own senses when necessary. Remember the experiment on how the brain reacted to chocolate, and the specific satiety that quickly emerged in the brain's reward center? The results are useful in everyday life. For those who suffer from a strong craving for unhealthy food, simply smelling it can actually help. If chocolate is your weakness, you can take a teaspoon of finely chopped chocolate wrapped in foil and smell it for five minutes when the craving sets in. The brain eventually gets tired of the smell, and the urge to eat a whole chocolate bar is reduced. In fact, sometimes just *thinking* about eating the food is enough to create a certain feeling of satiety.

The wine experts and master chefs show what an amazing world opens up when our sense of smell works with the rest of the brain. When we use our knowledge to refine our ability to enjoy food and drink. It's a world that doesn't actually require superhuman sensory abilities. It is accessible to us. Knowledge and experience are what allow food and drink to evoke rich emotions and experiences. Wine and food always have a story. The experience is not just a passive reaction to the chemicals and other sensory impressions you encounter, but to how your brain interprets them. According to the cognitive perspective, all the knowledge and associations you have acquired are transformed into preparations for new, exciting experiences of smell and taste. We do not experience food and drink with our mouth and nose, but with our brain. We need to remind ourselves of this. Companies spend a lot of resources on making food taste better and better. We should concentrate on making ourselves better tasters.

We are still in the age of sight where smells have been somewhat neglected. Earlier in this book I asked whether the superpowers of the sense of smell can be freed from the constraints of an image-obsessed culture. Will the future bring us a smell revolution? But along the way the answer has become clear to me. The sense of smell may have been sidelined, but smells have never lost their importance to us. They have continued to give us pleasure, interact with our personalities and guide our behavior through their ability to arouse hunger and thirst, attraction and disgust. After a long period in which the sense of smell has been both underestimated and downgraded by science, researchers are now rediscovering it as an interface where the different faculties of the brain meet and collaborate in a unique way. Every smell is an intersection between our thoughts and our emotions. A meeting place for the present, childhood and the beginning of life. Smelling encompasses all the other senses, and they are wordlessly mixed together. The sense of smell is the most ancient, but perhaps also the most refined of our senses. One aim of this book has been to make the reader *nosewise*: to realize that the sense of smell, often unnoticed, influences so many of the most important parts of our lives. Most of us have a world of smells to discover.

But this book will end with another, greater hope. It is about restoring the sense of smell to the millions of people who have lost it or suffered nightmarish distortions due to aggressive viral infections or other injuries. The sense of smell is fragile, but its vulnerability is also its strength. In the future, olfactory training methods, perhaps in combination with drugs that stimulate nerve growth, could restore what has been lost. We are not there yet, but the science of the sense of smell is advancing steadily. Restoring a damaged sense of smell would be the last stop on the long journey of rediscovering it—the journey that you are now part of. If that goal is achieved, we can safely say that we have, at last, entered the era of the sense of smell.

ACKNOWLEDGMENTS

I N THIS BOOK, I discuss many of the findings that I have encountered during my journey in research on the sense of smell. I've mentioned some of the key people behind that research, and I've referenced the most important sources for those who want to learn more. However, I am grateful to everyone who is or has been part of the chemosensory research community and who have advanced the study of the forgotten sense.

I would like to thank my Swedish and American publishers for their support. The English edition of the book was made possible thanks to HarperCollins Publishers and Mariner Books, who believed in my work, and I would especially like to thank Ivy Givens, who guided me through the process of editing the translated manuscript. The Swedish edition was supported by Natur & Kultur, and I would like to thank my editor Richard Herold and copyeditor Maria Sjödin. I also thank Lena Forssén, who encouraged me to write the book in the first place and who gave me valuable feedback throughout the process, as well as Magnus Linton, Patrik Hadenius and Arne Jarrick for giving me great advice on writing well and publishing.

Some of the chapters include content that was initially published as articles in the Swedish popular science magazines *Forskning & Framsteg, Modern Psykologi* and *Neurologi i Sverige,* as well as a chapter in the book *Ett kaleidoskop av kunskap* that I wrote together with historian Virginia Langum. A warm thanks to everyone who made it possible to publish these pieces.

My research is funded by the Knut and Alice Wallenberg Foundation via the Wallenberg Academy Fellows program, a consolidator grant from the Swedish Research Council and funds from the Swedish e-Science Research Centre. I wrote part of the book in the fall of 2022 at the Stellenbosch Institute for Advanced Study (STIAS) in South Africa. I thank the STIAS director, Edward K. Kirumira, the staff and the many wonderful colleagues that I was fortunate to get to know during this unforgettable visit. I learned so much from being able to conduct research in the laboratories of Jay Gottfried, Donald Wilson, Claire Murphy and John Polich, and I am thankful for their support. I am also grateful to all present and past members of my own research team, the Sensory Cognitive Interaction Laboratory, and my colleagues at Stockholm University for creating a stimulating and fun work environment. Several colleagues read and commented on parts of the text. A warm thanks to William Fredborg, Ingrid Ekström, Petter Kallioinen, Mikaela Pal, Teodor Jernsäther, Maria Larsson, Mats Olsson, Linus Andersson and Steven Nordin. The manuscript is much better thanks to your corrections and constructive suggestions. For all possible remaining errors and shortcomings in the final text, I am solely responsible.

Finally, a warm thanks to my wife, Caitlin Hawley, for her support, encouragement and patience. It means so much to me.

Stockholm, April 26, 2024
Jonas Olofsson

SOURCES

CHAPTER 1: THE SUPERPOWERS OF THE SENSE OF SMELL

Becher, P. G., Lebreton, S., Wallin, E. A., Hedenström, E., Borrero, F., Bengtsson, M., Joerger, V., & Witzgall, P. (2018). "The Scent of the Fly." *Journal of Chemical Ecology* 44(5), 431–435. https://doi.org/10.1007/s10886-018-0950-4.

Bushdid, C., Magnasco, M. O., Vosshall, L. B., & Keller, A. (2014). "Humans Can Discriminate More Than 1 Trillion Olfactory Stimuli." *Science* 343(6177), 1370–1372. https://doi.org/10.1126/science.1249168.

Enoch, J., McDonald, L., Jones, L., Jones, P. R., & Crabb, D. P. (2019). "Evaluating Whether Sight Is the Most Valued Sense." *JAMA Ophthalmology* 137(11), 1317–1320. doi:10.1001/jamaophthalmol.2019.3537.

Hansson, B. (2022). *Smelling to Survive: The Amazing World of Our Sense of Smell* (London: Legend Press Ltd.).

Herz, R. S., & Bajec, M. R. (2022). "Your Money or Your Sense of Smell? A Comparative Analysis of the Sensory and Psychological Value of Olfaction." *Brain Sciences* 12(3), 299. https://doi.org/10.3390/brainsci12030299.

Laska, M. (2017). "Human and Animal Olfactory Capabilities Compared." In A. Buettner (ed.), *Springer Handbook of Odor*, 675–690 (New York: Springer).

McCann Truth Central (2012). "Truth about Youth" (New York: McCann Worldgroup). https://issuu.com/mccanntruthcentral/docs/mccann_truth_about_youth.

McGann, J. P. (2017). "Poor Human Olfaction Is a 19th-Century Myth." *Science* 356(6338). eaam7263. https://doi.org/10.1126/science.aam7263.

Pellegrino, R., Hörberg, T., Olofsson, J., & Luckett, C. R. (2021). "Duality of Smell: Route-Dependent Effects on Olfactory Perception and Language." *Chemical Senses* 46, bjab025. https://doi.org/10.1093/chemse /bjab025.

Porter, J., Craven, B., Khan, R. M., Chang, S. J., Kang, I., Judkewitz, B., Volpe, J., Settles, G., & Sobel, N. (2007). "Mechanisms of Scent-Tracking in Humans." *Nature Neuroscience* 10(1), 27–29. https://doi.org/10.1038 /nn1819.

Sagan, C. (1979). *Broca's Brain: Reflections on the Romance of Science* (New York: Random House).

Schiller, F. (1992). *Paul Broca: Founder of French Anthropology, Explorer of the Brain* (New York: Oxford University Press).

Shepherd, G. (2013). *Neurogastronomy: How the Brain Creates Flavor and Why It Matters* (New York: Columbia University Press).

CHAPTER 2: CULTURAL CHEMICALS

Barwich, A. S. (2020). *Smellosophy: What the Nose Tells the Mind* (Boston: Harvard University Press).

Classen, C., Howes, D., & Synnott, A. (1994). *Aroma: The Cultural History of Smell* (New York: Routledge).

Farraia, M. V., Cavaleiro Rufo, J., Paciência, I., Mendes, F., Delgado, L., & Moreira, A. (2019). "The Electronic Nose Technology in Clinical Diagnosis: A Systematic Review." *Porto Biomedical Journal* 4(4), e42. https://doi.org/10.1097/j.pbj.0000000000000042.

Jacobs, L. F. (2019). "The Navigational Nose: A New Hypothesis for the Function of the Human External Pyramid." *Journal of Experimental Biology* 222(Pt Suppl 1), jeb186924. https://doi.org/10.1242/jeb.186924.

Johannisson, K. (2004). *Tecknen: Läkaren och Konsten att Läsa Kroppar* (Stockholm: Norstedts).

Kettler, A. (2020). *The Smell of Slavery: Olfactory Racism and the Transatlantic World* (New York: Cambridge University Press).

Muchembled, R. (2020) *Smells: A Cultural History of Odors in Early Modern Times* (Cambridge: Polity Press).

Murayama, R., & Swift, R. (2021). *Tasty TV:* "Japanese Professor Creates Flavorful Screen." Reuters. https://www.reuters.com/technology/lick -it-up-japan-professor-creates-tele-taste-tv-screen-2021-12-23/.

Niedenthal, S., Fredborg, W., Lundén, P., Ehrndal, M., & Olofsson, J. K. (2022). "A Graspable Olfactory Display for Virtual Reality." *International Journal of Human-Computer Studies* 169, *102928*.

Olofsson, J. K., & Langum, V. (2019). "Jag Luktar alltså Forskar Jag." In Håkansson (ed.), *Ett Kalejdoskop av Kunskap* (Stockholm: Santerus).

Palmer, R. (1993). "In Bad Odor: Smell and Its Significance in Medicine from Antiquity to the Seventeenth Century." In W. F. Bynum & Porter, R. (eds.), *Medicine and the Five Senses,* 61–68 (New York: Cambridge).

Turner, A., & Magan, N. (2004). "Electronic Noses and Disease Diagnostics." *Nature Reviews Microbiology* 2, 161–166. https://doi.org/10.1038/nrmicro823.

Wootton, D. (2016). *The Invention of Science: A New History of the Scientific Revolution* (London: Penguin Books).

CHAPTER 3: THE WORLD'S FIRST SMELLS

Buck, L., & Axel, R. (1991). "A Novel Multigene Family May Encode Odorant Receptors: A Molecular Basis for Odor Recognition." *Cell* 65(1), 175–187. https://doi.org/10.1016/0092-8674(91)90418-x.

Garcia-Burgos, D. (2021). "Contribution of Psychological Factors to the Affective Reactions toward Food Taste in Under- and Overnutrition." Lecture at European Chemoreception Research Organization, Cascais, Portugal.

Godfrey-Smith, P. (2018). *Other Minds: The Octopus and the Evolution of Intelligent Life* (London: HarperCollins UK).

Gisladottir, R. S., Ivarsdottir, E. V., Helgason, A., Jonsson, L., Hannesdottir, N. K., Rutsdottir, G., Arnadottir, G. A., Skuladottir, A., Jonsson, B. A., Norddahl, G. L., Ulfarsson, M. O., Helgason, H., Halldorsson, B. V., Nawaz, M. S., Tragante, V., Sveinbjornsson, G., Thorgeirsson, T., Oddsson, A., Kristjansson, R. P., Bjornsdottir, G., & Stefansson, K. (2020). "Sequence Variants in TAAR5 and Other Loci Affect Human Odor Perception and Naming." *Current Biology* 30(23), 4643–4653.e3. https://doi.org/10.1016/j.cub.2020.09.012.

Hansson, B. (2022). *Smelling to Survive: The Amazing World of Our Sense of Smell* (London: Legend Press Ltd.).

Herre, M., Goldman, O. V., Lu, T. C., Caballero-Vidal, G., Qi, Y., Gilbert, Z. N., Gong, Z., Morita, T., Rahiel, S., Ghaninia, M., Ignell, R., Matthews,

B. J., Li, H., Vosshall, L. B., & Younger, M. A. (2022). "Non-Canonical Odor Coding in the Mosquito. *Cell 185*(17), 3104–3123.e28. https://doi.org/10.1016/j.cell.2022.07.024.

Perszyk, E. E., Davis, X. S., Djordjevic, J., Jones-Gotman, M., Trinh, J., Hutelin, Z., Veldhuizen, M. G., Koban, L., Wager, T. D., Kober, H., & Small, D. M. (2023). "Odor Imagery but Not Perception Drives Risk for Food Cue Reactivity and Increased Adiposity." bioRxiv 2023.02.06.527292; doi: https://doi.org/10.1101/2023.02.06.527292.

Reed, D. (2021). "Twenty Years of Taste Receptors." Lecture at European Chemoreception Research Organization, Cascais, Portugal.

Spence, C. (2017). *Gastrophysics: The New Science of Eating* (New York: Viking).

Wagner, D. D., Dal Cin, S., Sargent, J. D., Kelley, W. M., & Heatherton, T. F. (2011). "Spontaneous Action Representation in Smokers When Watching Movie Characters Smoke." *Journal of Neuroscience* 31(3), 894–898. https://doi.org/10.1523/JNEUROSCI.5174-10.2011.

Wooding S. (2006). "Phenylthiocarbamide: A 75-Year Adventure in Genetics and Natural Selection." *Genetics* 172(4), 2015–2023. https://doi.org/10.1093/genetics/172.4.2015.

CHAPTER 4: THE EMOTIONAL TIME MACHINE

Arshamian, A., Iannilli, E., Gerber, J. C., Willander, J., Persson, J., Seo, H. S., Hummel, T., & Larsson, M. (2013). "The functional Neuroanatomy of Odor-Evoked Autobiographical Memories Cued by Odors and Words." *Neuropsychologia* 51(1), 123–131. https://doi.org/10.1016/j.neuropsychologia.2012.10.023.

Iatropoulos, G., Herman, P., Lansner, A., Karlgren, J., Larsson, M., & Olofsson, J. K. (2018). "The Language of Smell: Connecting Linguistic and Psychophysical Properties of Odor Descriptors." *Cognition* 178, 37–49. https://doi.org/10.1016/j.cognition.2018.05.007.

Larsson, M., Willander, J., Karlsson, K., & Arshamian, A. (2014). "Olfactory LOVER: Behavioral and Neural Correlates of Autobiographical Odor Memory." *Frontiers in Psychology* 5, 312. https://doi.org/10.3389/fpsyg.2014.00312.

Majid, A., & Burenhult, N. (2014). "Odors are Expressible in Language, as Long as You Speak the Right Language." *Cognition* 130(2), 266–270. https://doi.org/10.1016/j.cognition.2013.11.004.

Majid, A., Roberts, S. G., Cilissen, L., Emmorey, K., Nicodemus, B., O'Grady, L., Woll, B., LeLan, B., de Sousa, H., Cansler, B. L., Shayan, S., de Vos, C., Senft, G., Enfield, N. J., Razak, R. A., Fedden, S., Tufvesson, S., Dingemanse, M., Ozturk, O., Brown, P., & Levinson, S. C. (2018). "Differential Coding of Perception in the World's Languages." *Proceedings of the National Academy of Sciences of the United States of America* 115(45), 11369–11376. https://doi.org/10.1073/pnas.1720419115.

Olofsson, J. K., & Pierzchajlo, S. (2021). "Olfactory Language: Context Is Everything." *Trends in Cognitive Sciences* 25(6), 419–420. https://doi.org/10.1016/j.tics.2021.02.004.

Olofsson, J. K., & Gottfried, J. A. (2015). "The Muted Sense: Neurocognitive Limitations of Olfactory Language." *Trends in Cognitive Sciences* 19(6), 314–321. https://doi.org/10.1016/j.tics.2015.04.007.

Proust, M. (1913, 2003). *In Search of Lost Time: The Way by Swann's* (New York: Penguin Classics).

Schaal, B., Marlier, L., & Soussignan, R. (2000). "Human Fetuses Learn Odors from Their Pregnant Mother's Diet." *Chemical Senses* 25(6), 729–737. https://doi.org/10.1093/chemse/25.6.729.

Sorokowska, A., Sorokowski, P., & Frackowiak, T. (2015). "Determinants of Human Olfactory Performance: A Cross-Cultural Study." *Science of the Total Environment* 506-507, 196–200. https://doi.org/10.1016/j.scitotenv.2014.11.027.

Willander, J., & Larsson, M. (2007). "Olfaction and Emotion: The Case of Autobiographical Memory." *Memory and Cognition* 35(7), 1659–1663. https://doi.org/10.3758/bf03193499.

Zhou, G., Olofsson, J. K., Koubeissi, M. Z., Menelaou, G., Rosenow, J., Schuele, S. U., Xu, P., Voss, J. L., Lane, G., & Zelano, C. (2021). "Human Hippocampal Connectivity Is Stronger in Olfaction than Other Sensory Systems." *Progress in Neurobiology* 201, 102027. https://doi.org/10.1016/j.pneurobio.2021.102027.

CHAPTER 5: THE INTELLIGENCE OF THE NOSE

Andersson, L., Sandberg, P., Åström, E., Lillqvist, M., & Claeson, A. S. (2020). "Chemical Intolerance Is Associated with Altered Response Bias, Not Greater Sensory Sensitivity." *i-Perception* 11(6), 2041669520978424. https://doi.org/10.1177/2041669520978424.

Dantoft, T. M., Andersson, L., Nordin, S., & Skovbjerg, S. (2015). "Chemical Intolerance." *Current Rheumatology Reviews* 11(2), 167–184. https://doi.org /10.2174/1573397111021507021111101.

de Araujo, I. E., Rolls, E. T., Velazco, M. I., Margot, C., & Cayeux, I. (2005). "Cognitive Modulation of Olfactory Processing." *Neuron* 46(4), 671–679. https://doi.org/10.1016/j.neuron.2005.04.021.

Engen T. (1972). "The Effect of Expectation on Judgments of Odor." *Acta Psychologica* 36(6), 450–458. https://doi.org/10.1016/0001-6918(72)90025-x.

Hörberg, T., Larsson, M., Ekström, I., Sandöy, C., Lundén, P., & Olofsson, J. K. (2020). "Olfactory Influences on Visual Categorization: Behavioral and ERP Evidence." *Cerebral Cortex* 30(7), 4220–4237. https://doi.org /10.1093/cercor/bhaa050.

Johannisson, K. (2009). *Melankoliska Rum* (Stockholm: Albert Bonniers Förlag).

Köster, E.P., & de Wijk, R.A. (1991). "Olfactory Adaptation." In: D. G. Laing, R. L. Doty & W. Breipohl (eds.), *The Human Sense of Smell*, 199–215 (New York: Springer). https://doi.org/10.1007/978-3-642-76223-9_10.

Pellegrino, R., Hörberg, T., Olofsson, J., & Luckett, C. R. (2021). "Duality of Smell: Route-Dependent Effects on Olfactory Perception and Language." *Chemical Senses* 46, bjab025. https://doi.org/10.1093/chemse/bjab025.

Pierzchajlo, S., Jernsäther, T., Fontana, L., Almeida, R., & Olofsson, J. K. (2024). "Olfactory Categorization Is Shaped by a Transmodal Cortical Network for Evaluating Perceptual Predictions." *Journal of Neuroscience*, e1232232024 (advance online publication). https://doi.org/10.1523 /JNEUROSCI.1232-23.2024.

Porada, D. K., Regenbogen, C., Seubert, J., Freiherr, J., & Lundström, J. N. (2019). "Multisensory Enhancement of Odor Object Processing in Primary Olfactory Cortex." *Neuroscience* 418, 254–265. https://doi .org/10.1016/j.neuroscience.2019.08.040.

Small, D. M., Gerber, J. C., Mak, Y. E., & Hummel, T. (2005). "Differential Neural Responses Evoked by Orthonasal versus Retronasal Odorant Perception in Humans." *Neuron* 47(4), 593–605. https://doi.org/10.1016 /j.neuron.2005.07.022.

Small, D. M., Zatorre, R. J., Dagher, A., Evans, A. C., & Jones-Gotman, M. (2001). "Changes in Brain Activity Related to Eating Chocolate: From Pleasure to Aversion." *Brain* 124 (Pt 9), 1720–1733. https://doi .org/10.1093/brain/124.9.1720.

Slosson E. E. (1899). "A Lecture Experiment in Hallucinations." *Psychological Review* 6:407–408.

Zhou, G., Lane, G., Noto, T., Arabkheradmand, G., Gottfried, J. A., Schuele, S. U., Rosenow, J. M., Olofsson, J. K., Wilson, D. A., & Zelano, C. (2019). "Human Olfactory-Auditory Integration Requires Phase Synchrony between Sensory Cortices." *Nature Communications* 10(1), 1168. https://doi.org/10.1038/s41467-019-09091-3.

CHAPTER 6: FOLLOW YOUR NOSE

Ferdenzi, C., Roberts, S. C., Schirmer, A., Delplanque, S., Cekic, S., Porcherot, C., Cayeux, I., Sander, D., & Grandjean, D. (2013). "Variability of Affective Responses to Odors: Culture, Gender, and Olfactory Knowledge. *Chemical Senses* 38(2), 175–186. https://doi.org/10.1093/chemse/bjs083.

Gutiérrez, E. D., Dhurandhar, A., Keller, A., Meyer, P., & Cecchi, G. A. (2018). "Predicting Natural Language Descriptions of Mono-Molecular Odorants." *Nature Communications* 9(1), 4979. https://doi.org/10.1038/s41467-018-07439-9.

Haddad, R., Medhanie, A., Roth, Y., Harel, D., & Sobel, N. (2010). "Predicting Odor Pleasantness with an Electronic Nose." *PLoS Computational Biology* 6(4), e1000740. https://doi.org/10.1371/journal.pcbi.1000740.

Herz R. S. (2009). "Aromatherapy Facts and Fictions: A Scientific Analysis of Olfactory Effects on Mood, Physiology and Behavior." *International Journal of Neuroscience* 119(2), 263–290. https://doi.org/10.1080/00207450802333953.

Howard, S., & Hughes, B. M. (2008). "Expectancies, not Aroma, Explain Impact of Lavender Aromatherapy on Psychophysiological Indices of Relaxation in Young Healthy Women." *British Journal of Health Psychology* 13(Pt 4), 603–617. https://doi.org/10.1348/135910707X238734.

Hultén, B., Broweus, N., & Van Dijk, M. (2011). *Sinnesmarknadsföring* (Stockholm: Liber).

Spence, C. (2015). "Leading the Consumer by the Nose: On the Commercialization of Olfactory Design for the Food and Beverage Sector." *Flavor* 4, 31. https://doi.org/10.1186/s13411-015-0041-1.

CHAPTER 7: SMELLS AND INSTINCTS

Blomkvist, A. (2022). "Intimate Relationships and Olfaction: Body Odors, Adult Attachment, and Romance." (Doctoral dissertation, Stockholm University).

Brillat-Savarin, J-A. (1825, 2009). The Physiology of Taste (New York: Everyman's Library).

Doty, R. L. (2010). The Great Pheromone Myth (Baltimore: The Johns Hopkins University Press).

Dunbar, R. I. M. (2017). "Breaking Bread: The Functions of Social Eating." Adaptive Human Behavior and Physiology 3(3), 198–211. https://doi.org/10.1007/s40750-017-0061-4.

Granqvist, P., Vestbrant, K., Döllinger, L., Liuzza, M. T., Olsson, M. J., Blomkvist, A., & Lundström, J. N. (2019). "The Scent of Security: Odor of Romantic Partner Alters Subjective Discomfort and Autonomic Stress Responses in an Adult Attachment-Dependent Manner." Physiology & Behavior 198, 144–150. https://doi.org/10.1016/j.physbeh.2018.08.024.

Kamrava, S. K., Tavakol, Z., Talebi, A., Farhadi, M., Jalessi, M., Hosseini, S. F., Amini, E., Chen, B., Hummel, T., & Alizadeh, R. (2021). "A Study of Depression, Partnership and Sexual Satisfaction in Patients with Post-Traumatic Olfactory Disorders." Scientific Reports 11(1), 20218. https://doi.org/10.1038/s41598-021-99627-9.

Lundström, J. N., Mathe, A., Schaal, B., Frasnelli, J., Nitzsche, K., Gerber, J., & Hummel, T. (2013). "Maternal Status Regulates Cortical Responses to the Body Odor of Newborns." Frontiers in Psychology 4, 597. https://doi.org/10.3389/fpsyg.2013.00597.

Schaal, B., Coureaud, G., Langlois, D., Giniès, C., Sémon, E., & Perrier, G. (2003). "Chemical and Behavioral Characterization of the Rabbit Mammary Pheromone." Nature 424(6944), 68–72. https://doi.org/10.1038/nature01739.

Suskind, P. (2010). Perfume: The Story of a Murderer (London: Penguin Books Ltd.).

Stern, K., & McClintock, M. K. (1998). "Regulation of Ovulation by Human Pheromones." Nature 392(6672), 177–179. https://doi.org/10.1038/32408.

Wyatt T. D. (2020). "Reproducible Research into Human Chemical Communication by Cues and Pheromones: Learning from Psychology's Renaissance." Philosophical Tansactions of the Royal Society of London, series B, Biological Sciences 375(1800), 20190262. https://doi.org/10.1098/rstb.2019.0262.

CHAPTER 8: DIRTY SECRETS

Classen, C., Howes, D., & Synnott, A. (1994). Aroma: The Cultural History of Smell (New York: Routledge).

Croy, I., Nordin, S., & Hummel, T. (2014). "Olfactory Disorders and Quality of Life—An Updated Review." *Chemical Senses* 39(3), 185–194. https://doi.org/10.1093/chemse/bjt072.

Ferdenzi, C., Schirmer, A., Roberts, S. C., Delplanque, S., Porcherot, C., Cayeux, I., Velazco, M. I., Sander, D., Scherer, K. R., & Grandjean, D. (2011). "Affective Dimensions of Odor Perception: A Comparison between Swiss, British, and Singaporean Populations." *Emotion* 11(5), 1168–1181. https://doi.org/10.1037/a0022853.

Fialová, J. (2021). "The Effect of Diet on Human Body Odor Quality." Lecture at European Chemoreception Research Organization, Cascais, Portugal.

Herz, R. (2013). *That's Disgusting: Unraveling the Mysteries of Repulsion* (New York: W. W. Norton & Company).

Liuzza, M. T., Olofsson, J. K., Sabiniewicz, A., & Sorokowska, A. (2017). "Body Odor Trait Disgust Sensitivity Predicts Perception of Sweat Biosamples." *Chemical Senses* 42(6), 479–485. https://doi.org/10.1093/chemse/bjx026.

Liuzza, M. T., Lindholm, T., Hawley, C. B., Sendén, M., Ekström, I., Olsson, M. J., & Olofsson, J. K. (2018). "Body Odor Disgust Sensitivity Predicts Authoritarian Attitudes." *Royal Society Open Science* 5, *170191*.

Olsson, M. J., Lundström, J. N., Kimball, B. A., Gordon, A. R., Karshikoff, B., Hosseini, N., Sorjonen, K., Olgart Höglund, C., Solares, C., Soop, A., Axelsson, J., & Lekander, M. (2014). "The Scent of Disease: Human Body Odor Contains an Early Chemosensory Cue of Sickness." *Psychological Science* 25(3), 817–823. https://doi.org/10.1177/0956797613515681.

Kettler, A. (2020). *The Smell of Slavery: Olfactory Racism and the Transatlantic World* (New York: Cambridge University Press).

Lieberman, P., & Pizarro, D. (2010). "All Politics Is Olfactory." *New York Times.* https://www.nytimes.com/2010/10/24/opinion/24pizarro.html.

Zakrzewska, M. Z., Liuzza, M. T., Lindholm, T., Blomkvist, A., Larsson, M., & Olofsson, J. K. (2020). "An Overprotective Nose? Implicit Bias Is Positively Related to Individual Differences in Body Odor Disgust Sensitivity." *Frontiers in Psychology* 11, 301. https://doi.org/10.3389/fpsyg.2020.00301.

Zakrzewska, M. Z., Challma, S., Lindholm, T., Cancino-Montecinos, S., Olofsson, J. K., & Liuzza, M. T. (2023). "Body Odor Disgust Sensitivity Is Associated with Xenophobia: Evidence from Nine Countries across Five Continents." *Royal Society Open Science* 10(4), 221407. https://doi.org/10.1098/rsos.221407.

CHAPTER 9: SMELLS AND LIFE

Devanand, D. P. (2016). "Olfactory Identification Deficits, Cognitive Decline, and Dementia in Older Adults." *American Journal of Geriatric Psychiatry* 24(12), 1151–1157. https://doi.org/10.1016/j.jagp.2016.08.010.

Doty, R. L. (2012). "Olfactory Dysfunction in Parkinson Disease." *Nature Reviews Neurology* 8(6), 329–339. https://doi.org/10.1038/nrneurol.2012.80.

East, B. S., Fleming, G., Peng, K., Olofsson, J. K., Levy, E., Mathews, P. M., & Wilson, D. A. (2018). "Human Apolipoprotein E Genotype Differentially Affects Olfactory Behavior and Sensory Physiology in Mice." *Neuroscience* 380, 103–110. https://doi.org/10.1016/j.neuroscience.2018.04.009.

Ekström, I., Sjölund, S., Nordin, S., Nordin Adolfsson, A., Adolfsson, R., Nilsson, L. G., Larsson, M., & Olofsson, J. K. (2017). "Smell Loss Predicts Mortality Risk Regardless of Dementia Conversion." *Journal of the American Geriatrics Society* 65(6), 1238–1243. https://doi.org/10.1111/jgs.14770.

Fjaeldstad, A. W., & Smith, B. (2022). "The Effects of Olfactory Loss and Parosmia on Food and Cooking Habits, Sensory Awareness, and Quality of Life—A Possible Avenue for Regaining Enjoyment of Food." *Foods* 11(12), 1686. https://doi.org/10.3390/foods11121686.

Murphy C. (2019). "Olfactory and Other Sensory Impairments in Alzheimer Disease." *Nature Reviews Neurology* 15(1), 11–24. https://doi.org/10.1038/s41582-018-0097-5.

Olofsson, J. K., Ekström, I., Larsson, M., & Nordin, S. (2021). "Olfaction and Aging: A Review of the Current State of Research and Future Directions." *i-Perception* 12(3), 20416695211020331. https://doi.org/10.1177/20416695211020331.

Olofsson, J. K., Ekström, I., Lindström, J., Syrjänen, E., Stigsdotter-Neely, A., Nyberg, L., Jonsson, S., & Larsson, M. (2020). "Smell-Based Memory Training: Evidence of Olfactory Learning and Transfer to the Visual Domain." *Chemical Senses* 45, 593–600.

Olofsson, J. K., Nordin, S., Wiens, S., Hedner, M., Nilsson, L. G., & Larsson, M. (2010). "Odor Identification Impairment in Carriers of ApoE-varepsilon4 Is Independent of Clinical Dementia." *Neurobiology of Aging* 31(4), 567–577. https://doi.org/10.1016/j.neurobiolaging.2008.05.019.

Olofsson, J. K., Rönnlund, M., Nordin, S., Nyberg, L., Nilsson, L. G., & Larsson, M. (2009). "Odor Identification Deficit as a Predictor of Five-Year Global Cognitive Change: Interactive Effects with Age and ApoE-epsilon4." *Behavior Genetics* 39(5), 496–503. https://doi.org/10.1007/s10519-009-9289-5.

Pieniak, M., Oleszkiewicz, A., Avaro, V., Calegari, F., & Hummel, T. (2022). "Olfactory Training—Thirteen Years of Research Reviewed." *Neuroscience and Biobehavioral Reviews* 141, 104853. https://doi.org/10.1016/j.neubiorev.2022.104853.

Wegener, B-A., Croy, I., Hähner, A., & Hummel, T. (2018). "Olfactory Training with Older People." *International Journal of Geriatric Psychiatry* 33(1), 212–220. https://doi.org/10.1002/gps.4725.

CHAPTER 10: THE VIRUS THAT STOLE ALL THE SMELLS

Burges Watson, D. L., Campbell, M., Hopkins, C., Smith, B., Kelly, C., & Deary, V. (2021). "Altered Smell and Taste: Anosmia, Parosmia and the Impact of Long COVID-19." *PloS One* 16(9), e0256998. https://doi.org/10.1371/journal.pone.0256998.

Butowt, R., Bilinska, K., & von Bartheld, C. S. (2023). "Olfactory Dysfunction in COVID-19: New Insights into the Underlying Mechanisms." *Trends in Neurosciences* 46(1), 75–90. https://doi.org/10.1016/j.tins.2022.11.003.

de Melo, G. D., Lazarini, F., Levallois, S., Hautefort, C., Michel, V., Larrous, F., Verillaud, B., Aparicio, C., Wagner, S., Gheusi, G., Kergoat, L., Kornobis, E., Donati, F., Cokelaer, T., Hervochon, R., Madec, Y., Roze, E., Salmon, D., Bourhy, H., Lecuit, M., & Lledo, P. M. (2021). "COVID-19-Related Anosmia Is Associated with Viral Persistence and Inflammation in Human Olfactory Epithelium and Brain Infection in Hamsters." *Science Translational Medicine* 13(596), eabf8396. https://doi.org/10.1126/scitranslmed.abf8396.

Frasnelli, J., Tognetti, A., Winter, A., Thunell, E., Olsson, M.J., Greilert, N., Olofsson, J. K., Havervall, S., Thålin, C., & Lundström, J. N. (2024). "High Prevalence of Long-Term Olfactory Disorders in Healthcare Workers after Contracting COVID-19: A Case-Control Study." (unpublished manuscript).

Gerkin, R. C., Ohla, K., Veldhuizen, M. G., Joseph, P. V., Kelly, C. E., Bakke, A. J., Steele, K. E., Farruggia, M. C., Pellegrino, R., Pepino, M. Y., Bouysset, C., Soler, G. M., Pereda-Loth, V., Dibattista, M., Cooper, K. W., Croijmans, I., Di Pizio, A., Ozdener, M. H., Fjaeldstad, A. W., Lin, C., & GCCR Group Author (2021). "Recent Smell Loss Is the Best Predictor of COVID-19 Among Individuals with Recent Respiratory Symptoms." *Chemical Senses* 46, bjaa081. https://doi.org/10.1093/chemse/bjaa081.

Hummel, T., Rissom, K., Reden, J., Hähner, A., Weidenbecher, M., & Hüttenbrink, K. B. (2009). "Effects of Olfactory Training in Patients with Olfactory Loss." *Laryngoscope* 119(3), 496–499. https://doi.org/10.1002/lary.20101.

Kattar, N., Do, T. M., Unis, G. D., Migneron, M. R., Thomas, A. J., & McCoul, E. D. (2021). "Olfactory Training for Postviral Olfactory Dysfunction: Systematic Review and Meta-Analysis." *Otolaryngology—Head and Neck Surgery* 164(2), 244–254. https://doi.org/10.1177/0194599820943550.

Khan, M., Yoo, S. J., Clijsters, M., Backaert, W., Vanstapel, A., Speleman, K., Lietaer, C., Choi, S., Hether, T. D., Marcelis, L., Nam, A., Pan, L., Reeves, J. W., Van Bulck, P., Zhou, H., Bourgeois, M., Debaveye, Y., De Munter, P., Gunst, J., Jorissen, M., & Van Gerven, L. (2021). "Visualizing in Deceased COVID-19 Patients How SARS-CoV-2 Attacks the Respiratory and Olfactory Mucosae but Spares the Olfactory Bulb." *Cell* 184(24), 5932–5949.e15. https://doi.org/10.1016/j.cell.2021.10.027.

Juhlin, Richard. *"Världens bästa på Champagne."* *LoungePodden* 121. https://www.loungepodden.se/podcast/121-richard-juhlin-varldens-basta-pa-champagne/.

Niedenthal, S., Nilsson, J., Jernsäther, T., Cuartilles, D., Larsson, M., & Olofsson, J.K. (2021). "A Method for Computerized Olfactory Assessment and Training Outside of Laboratory and Clinical Settings." *i-Perception* 12.

Olofsson, J. K., Ekesten, F., & Nordin, S. (2022). "Olfactory Distortions in the General Population." *Scientific Reports* 12(1), 9776. https://doi.org/10.1038/s41598-022-13201-5.

Parma, V., Ohla, K., Veldhuizen, M. G., Niv, M. Y., Kelly, C. E., Bakke, A. J., Cooper, K. W., Bouysset, C., Pirastu, N., Dibattista, M., Kaur, R., Liuzza, M. T., Pepino, M. Y., Schöpf, V., Pereda-Loth, V., Olsson, S. B., Gerkin, R. C., Rohlfs Domínguez, P., Albayay, J., Farruggia, M. C., & Hayes, J. E. (2020). "More Than Smell—COVID-19 Is Associated with Severe Impairment of Smell, Taste, and Chemesthesis." *Chemical Senses* 45(7), 609–622. https://doi.org/10.1093/chemse/bjaa041.

Parker, J. K., Kelly, C. E., & Gane, S. B. (2022). "Insights into the Molecular Triggers of Parosmia Based on Gas Chromatography Olfactometry." *Communications Medicine* 2, 58. https://doi.org/10.1038/s43856-022-00112-9.

CHAPTER 11: BECOMING NOSEWISE

Avillez, J. (2021). "Food for Thought." Keynote lecture at European Chemo-reception Research Organization, Cascais, Portugal.

Ballester, J., Abdi, H., Langlois, J. *et al.* (2009). "The Odor of Colors: Can Wine Experts and Novices Distinguish the Odors of White, Red, and Rosé Wines?" *Chemosensory Perception* 2, 203–213. https://doi.org/10.1007/s12078-009-9058-0.

Bosker, B. (2017). *Cork Dork: A Wine-Fueled Journey into the Art and Science of Taste* (London: Allen & Unwin).

Hall, L., Johansson, P., Tärning, B., Sikström, S., & Deutgen, T. (2010). "Magic at the Marketplace: Choice Blindness for the Taste of Jam and the Smell of Tea." *Cognition* 117(1), 54–61. https://doi.org/10.1016/j.cognition.2010.06.010.

Kemps, E., & Tiggemann, M. (2007). "Modality-Specific Imagery Reduces Cravings for Food: An Application of the Elaborated Intrusion Theory of Desire to Food Craving." *Journal of Experimental Psychology Applied* 13(2), 95–104. https://doi.org/10.1037/1076-898X.13.2.95.

Morrot, G., Brochet, F., & Dubourdieu, D. (2001). "The Color of Odors." *Brain and Language*,7 9:309–320.

Shepherd, G. (2016). *Neuroenology: How the Brain Creates the Taste of Wine* (New York: Columbia University Press).

Spence, C. (2017). *Gastrophysics: The New Science of Eating* (New York: Viking).

White, T. L., Thomas-Danguin, T., Olofsson, J. K., Zucco, G. M., & Prescott, J. (2020). "Thought for Food: Cognitive Influences on Chemosensory Perceptions and Preferences." *Food Quality and Preference* 79, article 103776. https://doi.org/10.1016/j.foodqual.2019.103776.

INDEX